Yale Agrarian Studies Series
James C. Scott, Series Editor

The Agrarian Studies Series at Yale University Press seeks to publish outstanding and original interdisciplinary work on agriculture and rural society—for any period, in any location. Works of daring that question existing paradigms and fill abstract categories with the lived-experience of rural people are especially encouraged.
JAMES C. SCOTT, *Series Editor*

James C. Scott, *Seeing Like a State: How Certain Schemes to Improve the Human Condition Have Failed*

Michael Goldman, *Imperial Nature: The World Bank and Struggles for Social Justice in the Age of Globalization*

Steve Striffler, *Chicken: The Dangerous Transformation of America's Favorite Food*

Parker Shipton, *The Nature of Entrustment: Intimacy, Exchange, and the Sacred in Africa*

Alissa Hamilton, *Squeezed: What You Don't Know About Orange Juice*

Parker Shipton, *Mortgaging the Ancestors: Ideologies of Attachment in Africa*

Bill Winders, *The Politics of Food Supply: U.S. Agricultural Policy in the World Economy*

James C. Scott, *The Art of Not Being Governed: An Anarchist History of Upland Southeast Asia*

Stephen K. Wegren, *Land Reform in Russia: Institutional Design and Behavioral Responses*

Benjamin R. Cohen, *Notes from the Ground: Science, Soil, and Society in the American Countryside*

Parker Shipton, *Credit Between Cultures: Farmers, Financiers, and Misunderstanding in Africa*

Paul Sillitoe, *From Land to Mouth: The Agricultural "Economy" of the Wola of the New Guinea Highlands*

Sara M. Gregg, *Managing the Mountains: Land Use Planning, the New Deal, and the Creation of a Federal Landscape in Appalachia*

Michael R. Dove, *The Banana Tree at the Gate: A History of Marginal Peoples and Global Markets in Borneo*

Patrick Barron, Rachael Diprose, and Michael Woolcock, *Contesting Development: Participatory Projects and Local Conflict Dynamics in Indonesia*

Edwin C. Hagenstein, Sara M. Gregg, and Brian Donahue, eds., *American Georgics: Writings on Farming, Culture, and the Land*

Timothy Pachirat, *Every Twelve Seconds: Industrialized Slaughter and the Politics of Sight*

For a complete list of titles in the Yale Agrarian Studies Series, visit www.yalebooks.com.

Every Twelve Seconds

Industrialized Slaughter and the Politics of Sight

TIMOTHY PACHIRAT

Yale
UNIVERSITY PRESS
New Haven and London

Published with assistance from the foundation established in memory of Amasa Stone Mather of the Class of 1907, Yale College.

Yale University Press books may be purchased in quantity for educational, business, or promotional use. For information, please e-mail sales.press@yale.edu (U.S. office) or sales@yaleup.co.uk (U.K. office).

Set in Minion type by Integrated Publishing Solutions, Inc.
Printed in the United States of America.

The Library of Congress has cataloged the hardcover edition as follows:

Pachirat, Timothy, 1976–
Every twelve seconds : industrialized slaughter and the politics of sight / Timothy Pachirat.
p. cm. — (Yale agrarian studies series)
Includes bibliographical references and index.
ISBN 978-0-300-15267-8 (alk. paper)
1. Slaughtering and slaughter-houses—Social aspects—United States. 2. Meat industry and trade—Social aspects—United States. 3. Animal welfare—United States. 4. Pachirat, Timothy, 1976– I. Title.
TS1963.P33 2011
664′.9029—dc23 2011017018

ISBN 978-0-300-19248-3 (pbk.)

A catalogue record for this book is available from the British Library.

10 9 8 7 6 5 4

For Parker and Mia Jay-Pachirat,
and in memory of Jane Karen Pharnes Pachirat

"All right, but how do you get people to do the dirty work?" Oiie asked.

"What dirty work?" asked Oiie's wife, not following.

"Garbage collecting, grave digging," Oiie said.

Shevek added, "Mercury mining," and nearly said, "Shit processing," but recollected the Ioti taboo on scatological words. He had reflected, quite early in his stay on Urras, that the Urrasti lived among mountains of excrement, but never mentioned shit.

—Ursula Le Guin, *The Dispossessed*

"The problem with you is, you see too much and smell too much . . . ," said Avinash. "That's the secret—to distract your senses. Have I told you my theory about them? I think that our sight, smell, taste, touch, hearing are all calibrated for the enjoyment of a perfect world. But since the world is imperfect, we must put blinders on the senses."

—Rohinton Mistry, *A Fine Balance*

Slift (leaving): I advise you not to take up with those people down in the yards, they're a vile lot, frankly the scum of the earth.

Joan: I want to see it.

—Bertolt Brecht, *Saint Joan of the Stockyards*

Contents

Acknowledgments

Risking the maudlin, I want to begin by acknowledging the animals that died in the making of this book. Thirty-three million cattle are killed and sold as meat each year in the United States, and I aided in taking at least 240,000 of these creatures' lives in the nearly half a year I spent researching the work of killing. These cattle are part of the more than 8.5 billion animals that are slaughtered annually in the United States without respect or recognition, demonstrating the horrific efficiency of an industrialized food-production system that reduces sentient beings to raw material as well as the power of distance and concealment to make the unacceptable acceptable and the extraordinary ordinary. Although this book does not engage directly with arguments for animal rights, it is my deep hope that its detailed account of industrialized killing will invite readers to seek a more thoughtful relationship with the nonhuman creatures with whom we share the planet and a more critical stance toward the mechanisms of distance and concealment that currently operate to make those relationships palatable to the conscience as well as the stomach.

The violence of industrialized killing also cuts against humans, and I owe a great debt to my many co-workers on the kill floor. Without their patience, humor, and example, it is unlikely that I would have lasted more than a few days in the slaughterhouse. I am especially grateful to the workers of the cooler: Andrés, Carlos, Christian, Javier, Manuel, Ray, Tyler, Umberto, and, especially, Ramón, who drove me to work on many an early morning and kept me sane during interminable hours in the cooler. I am also grateful to my yellow-hat supervisor, Javier, and my red-hat supervisor, James: more than once, their seemingly trivial acts of kindness proved essential to my survival. I strive to tell the story of killing work in a way recognizable to everyone on the kill floor, including the kill floor managers and the U.S. Department of Agriculture inspectors. It is my hope that they too will find their experiences of killing work reflected in these pages.

The research for this book was possible only because Julie Jay and our daughters, Parker and Mia, agreed to live in Nebraska with me from 2004 to 2006. They tolerated my stench and my mustache (such as it was) while I worked on the kill floor and, I am happy to report, created thriving lives for themselves that had nothing to do with the killing of cattle. I am deeply thankful for the many moments away from the slaughterhouse that we spent together and for their ongoing encouragement and support throughout the writing of this book. I am also grateful to Sergio Sosa, Omaha Together One Community, our friends at COR, and the librarians at the University of Nebraska–Omaha, the University of Nebraska–Lincoln, and the Omaha Public Library for making our time in Omaha a rich one.

The intellectual soil for this project was tilled during an Agrarian Studies Seminar led by James C. Scott, Michael Dove, and Robert Harms; the seeds were planted in a seminar on Creativity and Methods, taught by James C. Scott and Arun

Agrawal; and it received abundant sunshine and rain at Yale University's Friday morning Agrarian Studies Colloquiums. James C. Scott offered enthusiastic support from the project's inception, and his intellectual inspiration and creative spirit animate this inquiry in big and small ways. Dvora Yanow has been a constant interlocutor on all matters methodological, honing my thinking and ethnographic sensibility. Seyla Benhabib, Pauline Jones-Luong, Ian Shapiro, David H. Smith, and Elisabeth Wood all stepped in with intellectual and bureaucratic aid at critical times. Steve Striffler, who worked undercover in a poultry plant in Arkansas and produced a first-rate account of it in *Chicken: The Dangerous Transformation of America's Favorite Food* (2005), provided invaluable advice and support before, during, and after my time on the kill floor.

Various chapters in this book have benefited enormously from participants at Edward Schatz's Political Ethnography Workshop at the University of Toronto, Dvora Yanow's Seminar in Organizational Ethnography at the Vrije Universiteit in Amsterdam, the Parsons–New School for Social Research Visual Culture Workshop, and, at Clarissa Hayward's invitation, the Political Theory Workshop on Politics, Ethics, and Society at Washington University in St. Louis. I would also like to thank students in my "Dirty and Dangerous Work," "Political Ethnography," and "M.A. Politics" seminars and the students of Hugh Raffles's "Politics of Nature" and Emanuele Castano's "Dehumanization" seminars at the New School for serving as classroom interlocutors on themes central to the book. Additionally, a courageous group of individuals—especially Julie Jay and Sathapon Pachirat—patiently worked their way through various iterations of the manuscript. Judith Grant, Nandini Deo, Lee Ann Fujii, Clarissa Hayward, Courtney Jung, Kathleen Letchford, Marvel Kay Mansfield, Monique Mironesco, and Kerri Willette provided more feedback than I could possibly take in: what I have neglected is to my own peril. At various times, Jessica Allina-Pisano, Jaskiran Dhillon, Vicky Hattam, Carrie Howerton, Mala Htun, Takeshi Ito, Robert and Mary Jay, Nomi Lazar, Lily Ling, Jim Miller, Richard Payne, Joy Roe-Pachirat, Melvin Rogers, Sanjay Ruparelia, Noy Thrupkaew, Dorian Warren, Lisa Wedeen, and Ashley Woodiwiss provided important friendship, intellectual companionship, advice, and support. At Yale University Press, Jean E. Thomson Black skillfully shepherded the overall project and Susan Laity brought a keen eye and reader's sensibility to copyediting the manuscript: an author could ask for no better advocates.

Bill Nelson and Jonathan Matthew Hoye took my hand-drawn fragments and transformed them into the kill floor maps that appear in Chapter 3. I also thank Tarek Masoud for his assistance with figure 9, and Asif Akhtar, Aleksandra Sekinger, Carlos Yescas, and Tomer Zeigerman for last-minute research and editorial assistance. The University of Chicago Press kindly provided permission to use material and passages from my chapter, "The Political in Political Ethnography: Dispatches from the Kill Floor," in *Political Ethnography: What Immersion Contributes to the Study of Power,* ed. Edward Schatz (Chicago: University of Chicago Press, 2009).

This book is dedicated to my daughters, Parker and Mia Jay-Pachirat, and to the memory of my mother, Jane Karen Pharnes Pachirat.

I
Hidden in Plain Sight

The slaughterhouse is cursed and quarantined
like a boat carrying cholera.
—*Georges Bataille*

In 2004, six cattle escaped from the holding pen of an industrialized slaughterhouse in Omaha, Nebraska. According to the *Omaha World Herald,* which featured the story on its front page, four of the six cattle made an immediate run for the parking lot of nearby Saint Francis of Assisi Catholic Church, where they were recaptured and transported back to be slaughtered. A fifth animal trotted down a main boulevard to the railroad yards that used to service Omaha's once-booming stockyards. The sixth, a cream-colored cow, accompanied the fifth animal partway before turning into an alleyway leading to another slaughterhouse.[1]

Workers from the first slaughterhouse and shotgun-armed Omaha police pursued the cream-colored cow into the alley, cornering it against a chain-link fence. After failing to herd the uncooperative cow into a waiting trailer, the police waved the workers back and opened fire on it. The cow ran a few steps, then fell, bellowing and struggling to rise while the police fired on it again.

The shooting took place during the ten-minute afternoon break for the workers at the second slaughterhouse. Venturing outside for fresh air, sunshine, and cigarettes, many of the slaughterhouse workers witnessed the killing of the animal firsthand, and during the lunch break the next day the news spread rapidly among the slaughterhouse employees, fueled by a graphic retelling by a quality-control worker who had been dispatched to the alleyway by slaughterhouse managers to observe the events and, later, to photograph the damage caused to the walls by errant shotgun pellets.

"They shot it, like, ten times," she said, her face livid with indignation, and her words sparked a heated lunch-table discussion about the injustice of the shooting and the ineptitude of the police. She began recounting the story of an unarmed man from Mexico who had recently been shot by the Omaha police. "They shot him just like they shot the cow," she asserted, to the nodding assent of her co-workers. "If he'd been white they wouldn't have shot him. You know, if you are Mexican in this country, the police will do anything to you."

I am driving south through the area of Omaha where the killing took place, and as I approach it a putrid odor, at once sharp and layered, seeps through the metal, rubber, and glass of my car, nestles in the cotton threads of my clothing, and forces a physical reaction that builds in my stomach and mouth before erupting acidly into my throat. I have experienced this sensation before, walking through the open-air northeastern Thai food markets of my childhood or driving by chocolate factories in New Jersey: smells so totalizing the nose sends them instantaneously to the tongue and plays them back as images in the mind.

As I exit the interstate, the odor intensifies. I am nearing the center of the industrialized slaughterhouse's olfactory kingdom. A roadside sign, erected by the city, reads, "To Report Manure Spills or Odor, Call 444-4919." An empty assertion of bureaucratic power over the unruliness of smell, it is one among numerous symptoms of the ongoing conflict between the messiness of mass killing and a society's—*our* society's—demand for a cheap, steady supply of physically and morally sterile meat fabricated under socially invisible conditions. Shit and smell: anomalous dangers to be reported to the authorities in an era in which meat comes into our homes antiseptically packaged in cellophane wrappings. To enable us to eat meat without the killers or the killing, without even—insofar as the smell, the manure, and the other components of organic life are concerned—the animals themselves: this is the logic that maps contemporary industrialized slaughterhouses, where in 2009 some 8,520,225,000 chickens, 245,768,000 turkeys, 113,600,000 pigs, 33,300,000 cattle, 22,767,000 ducks, 2,768,000 sheep and lambs, and 944,200 calves were killed for their meat in the United States.[2]

This book provides a firsthand account of contemporary, industrialized slaughter and does so to provoke reflection on how distance and concealment operate as mechanisms of power in modern society. Although we literally ingest its products in our everyday lives, the contemporary slaughterhouse is "a place that is no-place," physically hidden from sight by walls and socially veiled by the delegation of dirty, dangerous, and demeaning work to others tasked with carrying out the killing, skinning, and dismembering of living animals. Taking the contemporary slaughterhouse as an exemplary instance of how distance and concealment operate in our society, in this book I explore the work of industrialized killing from the perspective of those who carry it out, providing a close account of what it means to participate in the massive, routinized slaughter of animals for consumption by a larger society from which that work is hidden.[3]

Like its more self-evidently political analogues—the prison, the hospital, the nursing home, the psychiatric ward, the refugee camp, the detention center, the interrogation room, the execution chamber, the extermination camp—the modern industrialized slaughterhouse is a "zone of confinement," a "segregated and isolated territory," in the words of sociologist Zygmunt Bauman, "invisible" and "on the whole inaccessible to ordinary members of society." Close attention to how the work of industrialized killing is performed might thus illuminate not only how the realities of industrialized animal slaughter are made tolerable but the ways distance and concealment operate in analogous social processes: war executed by volunteer armies; the subcontracting of organized terror to mercenaries; and the violence underlying the manufacture of thousands of items and components we make contact with in our everyday lives. Such scrutiny makes it possible, as social theorist Pierre Bourdieu puts it, "to think in a completely astonished and disconcerted way about things [we] thought [we] had always understood."[4]

The physical escape of cattle from the Omaha slaughterhouse is also a conceptual escape, a rupture of categories.

Slaughtered by the tens of millions annually, six of these animals became front-page news when they briefly roamed freely through the city streets. Conceptually dangerous, their escape threatened to surface power relations that work precisely through confinement, segregation, and invisibility within a society that considers the manure—and even the smell—of these animals something to be reported to the authorities. In escaping the confines of the slaughterhouse, the cattle become, like the anthropologist Mary Douglas's definition of dirt, "matter out of place." And just as Douglas uses matter out of place to explore the taken-for-granted worlds of matter in place, so too does the escape of the Omaha cattle signal what might be learned about distance and concealment through a close exploration of the work of industrialized killing.[5]

Those who profit directly from contemporary slaughterhouses also actively seek to safeguard the distance and concealment that keep the work of industrialized killing hidden from larger society. On March 17, 2011, the Iowa State House of Representatives passed, by a vote of 66 to 27, HF 589, "A Bill for an Act Relating to Offenses Involving Agricultural Operations, and Providing Penalties and Remedies" (a similar bill is also under consideration in the Florida legislature). Supported by lobbyists for Monsanto, the Iowa Farm Bureau Federation, and the Iowa Cattlemen's, Pork Producers, Poultry, and Dairy Foods associations, the bill makes it a felony to gain access to and record what takes place in slaughterhouses and other animal and crop facilities without the consent of the facilities' owners. The broad scope and severe penalties of this attempt to further sequester industrialized killing and other contemporary practices of animal production from view are particularly highlighted in two sections of the bill, "Animal Facility

Interference" and "Animal Facility Fraud," which were explained in an earlier version of the bill, HF 431:

> INTERFERENCE. The bill prohibits a person from interfering with an animal facility. . . . This includes producing an audio or visual record which reproduces an image or sound occurring on or in the location, or possessing or distributing the record. It also prohibits a person from . . . entering onto the location, if the person has notice that the location is not open to the public. The severity of the offense is based on whether there has been a previous conviction. For the first conviction, the person is guilty of an aggravated misdemeanor, and for a second or subsequent conviction, the person is guilty of a class "D" felony.
>
> FRAUD. The bill prohibits a person from committing fraud, by obtaining access to an animal facility . . . by false pretenses for the purpose of committing an act not authorized by the owner, or making a false statement as part of an application to be employed at the location. The severity of the offense is based on whether there has been a previous conviction. For the first conviction, the person is guilty of an aggravated misdemeanor, and for a second or subsequent conviction, the person is guilty of a class "D" felony.[6]

The penalties for these offenses are severe:

> CONVICTION FOR OFFENSES—PENALTIES. A class "D" felony is punishable by confinement for no

> more than five years and a fine of at least $750 but not more than $7,500. An aggravated misdemeanor is punishable by confinement for no more than two years and a fine of at least $625 but not more than $6,250.
>
> CIVIL PENALTIES. In addition to the criminal penalties, a person suffering damages resulting from the commission of tampering or interference may bring an action in the district court against the person causing the damages to recover an amount equaling three times all actual and consequential damages, and court costs and reasonable attorney fees. In addition, a court may grant a petitioner equitable relief.[7]

The bill specifically criminalizes unauthorized physical access to industrialized slaughterhouses, unauthorized visual, audio, and print documentation of what takes place in slaughterhouses, and the possession and distribution of those unauthorized records regardless of who originally produced them.

A section of the bill detailing how the boundaries of the industrialized slaughterhouse and other animal production facilities are to be legally demarcated states: "A person has notice that an animal facility is not open to the public if the person is provided notice before entering onto the facility, or the person refuses to immediately leave the facility after being informed to leave. The notice may be in the form of a written or verbal communication by the owner, a fence or other enclosure designed to exclude intruders or contain animals, or a sign posted which is reasonably likely to come to the attention of an intruder and which indicates that entry is forbidden."[8]

Here, then, is a legal reinforcement of the industrialized

slaughterhouse's physical isolation. The fences and walls that quarantine the work of industrialized killing from larger society are specifically described in the bill as containing animals and excluding "intruders"; these physical barriers receive a special legal status that supersedes the legal status of other, less socially fraught fences and enclosures. What is more, "animal facility fraud" is invented as a new criminal category, applicable to those who seek employment in the industrialized slaughterhouse in order to reveal what takes place inside its walls. Like the physical walls of the slaughterhouse, slaughterhouse work is set apart as something that contains specific prohibitions and criminal sanctions inapplicable to more socially neutral forms of employment. Finally, the act of recording images and audio inside industrialized slaughterhouses as well as the mere possession and distribution of such recordings are criminalized, investing such images with a particular legal condemnation that sets them, too, apart from other images and audio recordings.[9]

The scope of the proposed bill and the severity of its penalties are indicators of the deep fear held by slaughterhouse owners and other financial beneficiaries of animal-production facilities about what might result if the work of industrialized killing and other contemporary animal-production practices were made visible. Much like the response provoked by the escaped Omaha cattle, its overt targeting of those who intentionally reveal what is hidden in plain sight signals the existence of power relations characterized by confinement, segregation, and invisibility.

An examination of the everyday realities of contemporary slaughterhouse work illuminates not only the ways in which the slaughterhouse is overtly segregated from society as a whole, but—paradoxically and perhaps more important—

how the work of killing is hidden even from those who participate directly in it. The workers who reacted with outrage and disgust to the shooting of a single cow by the Omaha police participate in the killing of more than 2,400 cattle on a daily basis. The immediacy of the killing by the police of one animal provoked a revulsion that is utterly absent in the day-to-day operations of the slaughterhouse, during which an animal is killed every twelve seconds. Distance and concealment shield, sequester, and neutralize the work of killing even, or especially, where it might be expected to be least hidden.

Exploring industrialized killing from this vantage point draws attention to the distance we create through walls, screens, catwalks, fences, security checkpoints, and geographic zones of isolation and confinement. It reveals the distance we create by constructing and reinforcing racial, gender, citizenship, and education hierarchies that coerce others into performing dangerous, demeaning, and violent tasks from which we directly benefit. It makes visible the distance we create with language—in the ways we avoid precise descriptions of repugnant things, inventing instead less dangerous names and phrases for them.[10] And, by employing a method of ethnographic immersion, it also uncovers the distance those who study the social world often create between themselves and the world(s) they claim the expertise to describe, analyze, and explain. In short, this is an account of industrialized killing that illuminates distance in four metrics: physical, social, linguistic, and methodological.

In attending to these metrics of distance, I engage two broad formulations about the relation between power and sight. The first, articulated by the historical sociologist Norbert Elias in his monumental work *The Civilizing Process*, posits "segregation, 'removing out of sight,' [and] concealment as the major method of the civilizing process." Tracing the dual processes of Western state formation and manners, Elias argues that concealment and the *creation* of distance mark the primary relation between power and sight in the contemporary era: "It will be seen again and again how characteristic of the whole process that we call civilization is this movement of segregation, this hiding 'behind the scenes' of what has become distasteful."[11]

Elias traces this broad movement in Western societies by demonstrating how, concurrent with the centralization of violence in the modern state, physical acts and states of being such as nudity, defecation, urinating, spitting, nose blowing, sexual intercourse, the killing of animals, and a host of others were increasingly identified as repugnant and removed from view. Drawing on Western etiquette manuals to document changes in public standards for bodily functions, nakedness, sexual relations, table manners, attitudes toward children, and the treatment of animals from the sixteenth through the nineteenth century, Elias convincingly reveals the following pattern: what once occurred in the open without provoking reactions of either moral or physical disgust has been increasingly segregated, confined, and hidden from sight. Manners surrounding the eating of meat are identified as particular historical evidence: table portions grow smaller, making meat less identifiably animal. "Carving knives also shrink, all the less to recall the instrument that deals the death stroke. . . . Reminders that the meat dish has something to do with the killing of an animal are avoided to the utmost. In many of our meat dishes the animal form is so concealed and changed by the art of its preparation and carving that, while eating, one is scarcely reminded of its origin."[12]

"Civilization," which commonly presents itself to those

living in contemporary industrialized societies and urban areas as a ready-made product suitable for inculcation in children and "barbarians," is in fact a long historical process still in the making, the political implications of which have yet to be fully understood. Key to an understanding of these implications is an exploration of what it means that a central characteristic of what are referred to as development and progress relies on the distancing and concealment of morally and physically repugnant practices rather than their elimination or transformation.

The account of industrialized slaughterhouse work in this book offers a detailed exploration of precisely the kind of phenomenon identified by Elias: a labor considered morally and physically repellent by the vast majority of society that is sequestered from view rather than eliminated or transformed. Considering this hidden work from the standpoint of those who perform it, however, also makes relevant an alternative formulation about the relation between power and sight that stands in contrast to Elias's emphasis on segregation and confinement. In this alternative formulation, a central mechanism of power in the contemporary era works by *removing* barriers to sight, by eradicating obstacles that create possibilities for darkness and concealment, and by installing instead what the social theorist Michel Foucault identified as "continuous and permanent systems of surveillance."[13]

Drawing on Jeremy Bentham's architectural plan for a new kind of prison, which he called the Panopticon, Foucault outlines how visibility functions as a mechanism of power:

> The principle was this. A perimeter building in the form of a ring. At the centre of this, a tower, pierced by large windows opening on to the inner face of

> the ring. The outer building is divided into cells each of which traverses the whole thickness of the building. These cells have two windows, one opening on to the inside, facing the windows of the central tower, the other, outer one allowing daylight to pass through the whole cell. All that is then needed is to put an overseer in the tower and place in each of the cells a lunatic, a patient, a convict, a worker or a schoolboy. The back lighting enables one to pick out from the central tower the little captive silhouettes in the rings of cells. In short, the principle of the dungeon is reversed; daylight and the overseer's gaze capture the inmate more effectively than darkness, which afforded after all a sort of protection.

In Bentham's proposal for prison reform, surveillance—internalized by the prisoners to the point where they would police themselves—would replace overt physical punishment as the dominant mechanism of control over individuals. For Foucault, this ideal of total visibility underlies the application of modern disciplinary power across a variety of settings: prisons, insane asylums, military barracks, schools, and factories. It is a mechanism of power in which all is brought to light and nothing is hidden: "In the Panopticon each person, depending on his place, is watched by all or certain of the others. You have an apparatus of total and circulating mistrust, because there is no absolute point. The perfected form of surveillance consists in a summation of malveillance."[14]

In an analysis of sight and power from the perspective of the state rather than the specific architectural instance of the Panopticon, James C. Scott also identifies the desire for increased visibility as a hallmark of the operation of power. Whether it be trees or people, Scott argues, a central characteristic of modern power structures is their impulse to rearrange and, if necessary, exterminate and create anew their subjects in ways that approximate an ideal of perfect visibility. This visibility, in turn, serves fantastical and fanatical projects of control, often under the legitimizing rhetoric of improvement and development of the very populations being fit into the grid. Mixed-growth forests are replaced with trees planted in straight lines conducive to counting and cutting; intercropping gives way to industrialized monocropping; ambiguous loyalties in overlapping systems of authority yield to clear-cut national borders and citizenship categories; and nomadic peoples are fixed in place and assigned surnames for the purposes of taxation, control, and "development." As with the Panopticon, this is a logic of power directly linked to an *expansion* of sight, a *leveling* of obstacles to visibility and transparency.[15]

Scott's state-centric perspective is later reversed when he focuses on the pre-1950s history of non-state spaces in the Southeast Asian massif, an area he terms Zomia. Richly describing the state-repelling geographic, agricultural, cultural, and linguistic technologies and tactics that constituted and defended one of the largest continuous non-state spaces in the world from encroachment by the rice-paddy state, Scott nonetheless concludes that these spaces are all but extinct, overrun by postcolonial lowland states employing a variety of strategies and tactics that materially share a powerful arsenal of what he terms "distance-demolishing technologies." The list of these technologies includes all-weather roads, bridges, railroads, modern weapons, telegraph, telephone, airpower, helicopters, and modern information technologies, such as global navigation satellite systems.[16]

These technologies are an extensification of the Panopticon: they work to expand the range of vision of a controlling overseer and, with that expansion, to come closer to realizing the fantasy of total transparency, the banishment of concealment. Power structures work here by "demolishing distance," both the distance that prevents the creation of self-policing individuals who have internalized an external gaze (in Foucault's disciplinary power) and the distance that depends on altitude, rugged terrain, and the cultivation of root crops to repel advances by lowland, labor-intensive rice kingdoms (in Scott's state power). The overall arch of this alternative formulation of the relation between sight and power is unmistakable: power operates by collapsing distances and exposing concealed spaces.

How might these broad characterizations about the relation of modern power to sight be understood together? One advances the idea that power operates through the creation of distance and concealment and that our understandings of "progress" and "civilization" are inseparable from, and perhaps even synonymous with, the concealment (but not elimination) of what is increasingly rendered physically and morally repugnant. Its alternative counters that power operates by collapsing distance, by making visible what is concealed.

The account of industrialized slaughter offered in these pages demonstrates how these seemingly contradictory characterizations relate in practice. By concentrating their vast historical sweep in an examination of what it means to actually carry out the work of contemporary killing in a society that hides such work in plain sight, I show how surveillance and concealment work together, how quarantine is possible in, and perhaps even enabled by, conditions of total visibility. Attention to industrialized killing from the vantage point of those who perform it demonstrates the capacity for sequestration and surveillance to exist in symbiosis as mechanisms of power in contemporary society. And as I explore in the book's final chapter, the potential for this symbiosis carries implications for movements from across the political spectrum that engage in what I term a politics of sight, defined as organized, concerted attempts to make visible what is hidden and to breach, literally or figuratively, zones of confinement in order to bring about social and political transformation.

This book also, of course, itself enacts a politics of sight by making visible a massive, routinized work of killing that many would prefer to keep hidden. It breaches the zone of confinement that is industrialized slaughter, challenging both those in broader society who want to consume the products of the slaughterhouse while keeping its realities hidden from view and those who financially profit from—and therefore seek to criminalize any unauthorized revelations about—the practices of contemporary industrialized killing. To make these practices visible from the perspective of those who perform them, I participated directly in the work of industrialized killing, gaining full-time employment in a slaughterhouse in Omaha from June through December 2004. During these five and a half months, I worked full-time on the kill floor, Monday through Friday, for nine to twelve hours a day, starting between 5:00 and 7:00 A.M. and finishing between 4:00 and 6:30 P.M.

Seeking employment as an entry-level worker without informing the management that I intended to write about my experiences, I started out as a liver hanger in the cooler at $8.50 an hour; moved to the chutes, where I drove live cattle into the knocking box to be shot; and finally was promoted to quality control, a position that paid $9.50 an hour and gave me access to almost every part of the kill floor. Serendipity and improvisation governed my movement through these different positions in the slaughterhouse, but given my interest in how industrialized killing might generate insights into the operation of distance and concealment in society at large, I could not have chosen three better jobs. Liver hanging in the frigid cooler put me at maximal distance from the killing, chute work put me into personal contact with the live animals and their slaughter, and my quality-control work brought me directly into the internal hierarchies of the slaughterhouse and made me a participant in its adversarial relationships with the U.S. Department of Agriculture inspectors.[17]

My movement from one position at the slaughterhouse to another structured not only what I saw but how I saw it and how I gave meaning to it. Once inside as an active participant, I found myself inextricably caught up in its networks of power, its "webs of local associations." In addition to the assumptions that surrounded my various jobs in the slaughterhouse, my self-presentation, appearance, and mannerisms combined to create certain interpretations of me by others. Primary, perhaps, was my appearance: the son of one Southeast Asian and one white parent, I have dark-brown skin, black hair, and narrow brown eyes. In the employment trailer, these features helped me get hired. On the kill floor, they were often misread: many co-workers were incredulous to learn that I was not Mexican; still others could not understand that I was Asian but not Chinese or Vietnamese. My bilingual English and Thai, my halting Spanish, and the relative confidence, trained through years of formal education, with which I voiced opinions and asked questions both impeded and facilitated my interactions with co-workers, supervisors, and USDA inspectors. Not least, I was a male in a male-dominated workplace, which made it extremely difficult for me to form relationships with the twelve or so females who worked on the kill floor. All these factors and more affected how I was seen by others and, consequently, what I was able to see.[18]

In addition to my hours working on the kill floor, I also spent time with slaughterhouse workers and USDA inspectors outside work, informing them in each instance of my intention of writing a book and obtaining their consent to use information they provided. In December 2004, finding the ethical dilemmas involved with my quality-control work untenable, I resigned and left the slaughterhouse, after which I spent an additional year and a half in Omaha conducting interviews with slaughterhouse workers and assisting community organizing groups on slaughterhouse-related issues.

I entered the kill floor to provide an account of contemporary industrialized slaughter, not to expose a specific place. Had the latter been my goal, I would have had ample opportunity when, in a catalyst for my resignation, a USDA inspector approached me and asked me to testify about my knowledge of food-safety practices in the slaughterhouse. Based on my commitment not to implicate specific individuals or places in my research, I declined to testify, and the slaughterhouse I worked in remains unnamed in this book. Likewise, most individual names have been changed.[19]

The slaughterhouse I worked in continues to operate today. It employs close to eight hundred nonunionized workers, the vast majority immigrants and refugees from Central and South America, Southeast Asia, and East Africa. It generates over $820 million annually in sales to distributors within and outside the United States and ranks among the top handful of U.S. cattle-slaughtering and beef-processing facilities in volume of production. The line speed on the kill floor is approximately three hundred cattle per hour. In a typical workday, between twenty-two and twenty-five hundred cattle are killed there, adding up to well over ten thousand cattle killed per five-day week, or more than half a million cattle slaughtered each year.

In this book I employ a narrative format, often quoting conversations verbatim and letting the sensory, corporeal complexity of the slaughterhouse take precedence over neatly hewed analytical insights. My account relies centrally on context, with an emphasis on little things and multiple voices, and with a tolerance for ambiguity: I strive, in short, for "a writing strategy in which curiosity is not overwhelmed by coherence."[20]

This narrative format reverses a long-standing tradition in academic writing in which a deductive, often linear analytical argument structures the writing; in this tradition, if ethnographic fieldwork notes or verbatim quotations from conversations make an appearance at all, they do so as docile, heavily policed excerpts. Typically, these truncated descriptions or conveniently supportive quotations from informants are strategically sprinkled throughout the text to bolster both the analytic argument and the ethnographic authority of the author. By aiming for a writing strategy molded largely by the requirements of narrative rather than analysis, I hope to make room for the ambiguities, silences, and multiplicities in the experience of the work of killing. This strategy also challenges the reader to use these narratives as a way to think through what it means, from the perspective of lived experience, to perform the daily work of industrialized killing.

"My whole work has come to resemble a terrain of which I have made a thorough, geodetic survey, not from a desk with pen and ruler, but by touch, by getting down on all fours, on my stomach, and crawling over the ground inch by inch, and this over an endless period of time in all conditions of weather," wrote Henry Miller in "Reflections on Writing." In the pages that follow, I abandon desk and ruler and provide a close-range account of the daily work of industrialized killing, of what it means to smell, see, hear, taste, and touch it. This account is sensory but not intentionally sensational, not merely another contribution to an "anthropology that seeks out the loathsome and disgusting and delights in it," in the words of Ian Miller. After all, the persistent, dull ache of a wet glove against a bare hand in the near-freezing slaughterhouse cooler and the pen scratching hurriedly over the mandatory food-safety paperwork in the quality-control office are as much a part of what it means to perform the work of modern industrialized killing as the cutting smell of diarrhea in the chutes and the soft, mechanical *pffft-pffft* of the captive-bolt gun penetrating the skulls of steers.[21]

You may find the descriptions in the pages ahead both physically and morally repugnant. Recognize, however, that this reaction of disgust, this impulse to thumb through the pages so as to locate, separate, and segregate the sterile, abstract arguments from the flat, ugly, day-in, day-out minutiae of the work of killing, is the same impulse that isolates the slaughterhouse from society as a whole and, indeed, that sequesters and neutralizes the work of killing even for those who work within the slaughterhouse itself. The detailed accounts that follow are not merely incidental to or illustrative of a more important theoretical argument about how distance and concealment operate as mechanisms of power in contemporary society. They *are* the argument.

II
The Place Where Blood Flows

Job Number 114, Toenail Clipper: *uses hands to stick each cooked foot into machine that operates by two serrated rollers. The rollers, activated by foot lever, spin together and break off "toenail," or tip of hoof, of each cow foot.*

The wind grants wings to the spilled blood of the slaughterhouse, carrying its stench to the farthest reaches of the city in one direction before halting, reversing, and invading the opposite end. An old-timer living five miles east of the Omaha stockyards on the west bank of the Missouri River remembers the wind pushing the odors of the dead and dying through the walls of her family's home, subjugating them for days at a time to the olfactory reign of the slaughterhouse. In the mid-twentieth century, Omaha surpassed even Chicago, Carl Sandburg's "hog butcher for the world" and the city of Upton Sinclair's *The Jungle,* in livestock sales. Workers for Armour, Cudahy, Swift, and Wilson killed almost six and a half million cattle, hogs, and sheep in South Omaha each year. "It was the smell of money," the old-timer said of the stench from that killing; "it was the smell of money."[1]

At the turn of the twenty-first century, the skeletal remains of South Omaha's meatpacking prominence were still visible: an intricate two hundred–acre origami of corrals folded in on themselves in pen after pen, pale-gray wood wrinkled by decades of prairie sun, wind, and snow, as well as by the continuous, patient erosion of saliva, urine, and feces and the friction of the hides of millions of creatures driven through them to be traded and slaughtered. Shortly after the year 2000, the last corrals were razed to the ground, and they survive only in black-and-white photographs mounted on the walls of Omaha's steakhouses, community colleges, and labor-union halls. The old stockyard area is now a sprawling complex of big-box buildings gathered together in an industrial and commercial business park. Only the 1926 red-brown brick Livestock Exchange Building remains, converted into eleven stories of apartments that serve as an anachronistic monument to the complex of pens and corrals that once emanated from it.[2]

Today a casual passerby could be forgiven the assumption that the stockyard and slaughter industry have left South Omaha completely, replaced by what is readily visible: a tool manufacturing plant, a bank with three drive-through teller lanes, a big-box pet-supply store with a veterinarian clinic, a pawn shop with bicycles and lawnmowers out front, a branch of the Omaha Metropolitan Community College, a Chinese restaurant, a shoe store, a grocery store, and a video rental store. But the eyes are easy to deceive, the nose less so, and my search for the origin of the miasma that continues to pervade South Omaha despite its makeover is rewarded when the road undulates beneath my car. The movement is caused by a truck pulling a trailer whose metal sides are pitted with oval holes about the size of my outstretched hand. Peering through my car window into the darkness beyond one of the holes, I barely make out a shimmer, something catching the sunshine and throwing it back. I stare, puzzled, and slowly realize that it is the eye of a cow.

The truck turns into a smaller side street behind a large boxlike building surrounded by an expansive, car-filled asphalt parking lot enclosed by a chain-link fence topped with coiled barbed wire. The truck stops with a hiss in front of a small white rectangular structure, and an inky-haired woman in a dark-blue uniform and fluorescent orange vest with "SECURITY" printed on it in bold black letters across a silver reflective strip steps out and exchanges words with the truck driver. A moment passes, she waves her hand, and the truck hisses again, rolls forward, and turns in a wide arc before backing up in fits and starts to a concrete ledge. The cab door opens and two brown calf-high rubber boots fly out, hit the parking lot with a slap, bounce up, then hit the ground again, flat on their sides. A large bearded white man steps out, stretches, and puts one foot onto the silver metal step of his cab. He loosens a shoelace, takes off his boot, and uses his sock-clad toe to pull one of the rubber boots into an upright position before slipping his leg into it. He repeats the movement with the other foot, then walks to the back of the truck. A loud metallic thud is followed by three more, each less loud and more closely spaced than the one before it. A human voice shouts. The trailer sways. A rhythmic clang starts up, hesitant, then faster and more insistent. It crescendos: the beat of a hundred hoofs on metal. I have found the place where blood flows.

Circumnavigating the road that follows the perimeter of the fence, I am astonished. I could not have precisely articulated what I had subconsciously anticipated at a modern industrialized slaughterhouse: perhaps not quite rivulets of blood running through uncovered drains, screams of distressed animals, decomposing kidneys and lungs floating in open cesspools of feces, or muscled white-clad butchers drenched in blood, strutting about with gargantuan cleavers—but nothing in my imagination had prepared me for the utter invisibility of the slaughter, the banal insidiousness of what hides in plain sight. Facing outward, this industrialized slaughterhouse blends seamlessly into the landscape of generic business parks ubiquitous to Everyplace, U.S.A., in the early twenty-first century.

This city block–wide windowless box of gray corrugated steel set on a shoulder-high slab of concrete, topped by massive ceiling fans, surrounded by black asphalt, a chain-link fence, and guard huts, and fronted by a modern office complex of glass and aluminum presents only the generic face of mass production. The building materials, size, structure, and layout of the slaughterhouse could pass for the community college to the south, the tool factory to the east, or the pet-supply store to the north. Only the coming and going of hole-pocked semitrailers, the rhythmic clanging of hoofs, and the omnipresent stench intimate what lies inside.

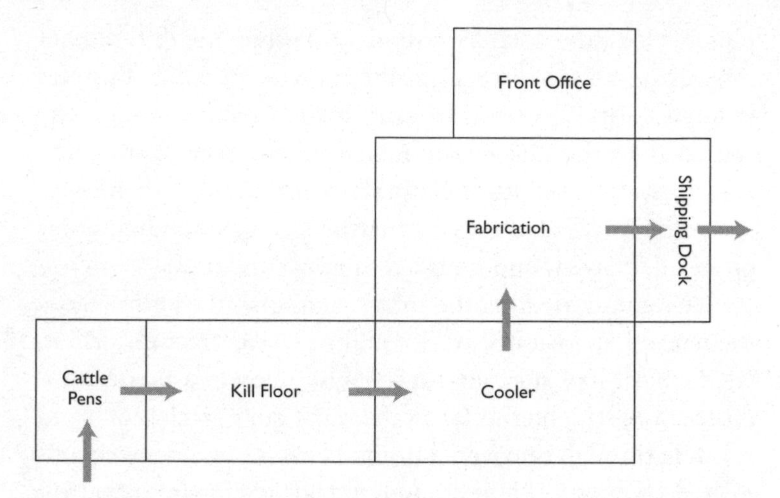

Fig. 1 Departmental Divisions of an Industrialized Slaughterhouse

The Front Office

A big, square, stern man
Sat in a swivel chair of oak
Beyond the door, the air stank putrid
Upon the nostrils of hurrying, garrulous,
Gesticulating stockmen.
The milling feet
Of many thousand cows, bulls, and shrill-voiced calves,
The frightened bleats
Of myriad lambs and dirty ewes,
The grunts of swine
The oaths of the men and the brays of the mules,
Who were building his dream—
Made music
For the big man's ears.[3]

Visiting outsiders—prospective customers, government officials, and employees of nearby medical or ophthalmology schools collecting organs or eyeballs for dissection—are welcomed to the front office, a modern-looking office building that slumps like a misplaced appendage in front of a windowless, towering, corrugated-metal structure topped by a rectangular block of whirring fans enclosed in gray metal cases. Concrete steps lead from the parking lot up to dark-glass double doors that bisect the office building into an east and west side. The west half is an impressive two-story wall of green glass framed by large vertical columns of shiny steel. The east half consists of a single story of concrete wall bisected by a narrow strip of the same green glass that fronts the west side. The contrast suggests conflict: the west end projects power and transparency; the east end, a bunkerlike mentality of secrecy and siege. And looming directly behind them, the enormous corrugated-metal block of the slaughterhouse overshadows this conflict—slumped diminutively against it, both halves of the office are reduced literally to their dramaturgical role: a sanitized front for the dirty work within (fig. 1).

Entering through the front doors of the office, an outsider encounters a white female receptionist who sits behind a desk across from two plush leather chairs. Between the chairs a small coffee table holds a stack of meatpacking-industry magazines, each showcasing an accomplishment of the company. Behind them silver and gold plaques for workplace safety adorn the walls. The receptionist's desk is made of a wide plank of dark wood that extends past its base and hangs stylishly out into the empty air. There is no partition behind the desk, allowing an unbroken line of sight to the eastern wall of the room, itself framed by a large window overlooking the manicured lawn surrounding three sides of the front office.

A long open worktable dotted with flat-screen computers, six to a side, occupies the space between the receptionist and the eastern wall. Twelve white, crew-cutted salesmen in dress pants, long-sleeved shirts, and ties, ranging in age from their mid-thirties to early fifties, sit in leather office chairs in front of these screens, each using a handless telephone set. Their computer screens show spreadsheets, graphs, maps, the Drudge Report, and the results of a Google search. South of this workspace, the room narrows into a hallway framed by the opaque walls of a vice president's office on the east and a conference room to the west. The hallway itself terminates at a closed door of galvanized steel.

Of the conference room's four walls, three are transparent glass, supported at five-foot intervals by black steel beams that meet the floor at an angle, rising outward to the ceiling. Through these glass walls, the visiting outsider sees a shiny metallic conference table encircled by more leather chairs. Soft spotlights embedded in the ceiling cast an uneven yellow light across the silvery table. The fourth, opaque wall is plain, with only a small window, no larger than a few feet square, in the center. Shades operated from within the conference room hang over the window.

West of the conference room, the front office space opens into a two-story room from which a dazzle of aquamarine light streams in through its translucent north wall. Immediately behind the wall, chest-high gray fabric and black plastic dividers partition the floor space into cubicles occupied by administrative and clerical workers and a quality-assurance and food-safety manager, all white women. West of these cubicles, the space terminates in a ten inch–thick pocket door constructed of solid metal. Behind this door, at the northwest corner of the office building—in an area maximally distant from the slaughterhouse's zones of production—is the office of the president, a white man.

South of the cubicles and the president's office, braided metal cables hang in regular gradations from the ceiling, supporting steps of poured concrete that lead up to a narrow overhang. A life-sized fiberglass cow painted with abstract red and black designs stands under the stairs, its eyes staring blankly ahead. Behind the stairway, under the overhang, wooden partitions demarcate bathrooms and areas for a photocopier, a fax machine, mailboxes for the front-office employees, and an alcove with a small table, refrigerator, microwave, and electric drip-coffee machine. Overhead, the overhang houses two glass-fronted offices belonging to the human-resources manager and the second vice president, both white women. To the west, these offices dead end into a frosted-glass door that opens into a large conference room with a long rectangular glass table surrounded by still more leather chairs.

This collection of flat-screen computers, conference rooms, and cubicle partitions; this assemblage of white- and pink-collar workspaces with their industry magazines, award plaques, and coffee machine; this domain of internal memos, button-down shirts, and phone extensions; this facsimile of countless other contemporary office spaces in metropolises and suburbs across the world is a sterilized bubble, an anemic appendage to the massive slaughterhouse, with its muscular sphere of production. The efforts at distinction, so notable in the tended lawns surrounding the office on three sides, are betrayed by the metal wall that severs it from the rest of the slaughterhouse, at once marking the southern boundary of the front office and towering above it. This wall both demarcates and enables the volatile combinations of citizenship, race, class, and education that separate the industrialized slaughterhouse's zones of privilege from its zones of production.

The wall divides categories as well as physical space: front office versus back room; task-conceiving versus task-executing;

creative versus rote; managerial versus subordinate; north versus south; white versus brown, yellow, and black; clean versus dirty; "civilized" versus "barbarian." Of the slaughterhouse's more than eight hundred employees, no more than twenty-five work in the front office, sequestered and protected by that wall, which authorizes them to manage decisions and processes that determine much of the texture of life for those who labor behind it. In this sense, the front office of the slaughterhouse is indistinguishable from front offices worldwide in which the control of others' lives is directed from a distance, opaquely, without the benefit (or the burden) of visceral, direct experience. If aerial bombardments—and unmanned U.S. Air Force drones in particular—serve as an exemplar of this kind of distancing at its most technologically supreme, the brute, physical materiality of the southern wall of the front office of the industrialized slaughterhouse offers a stark reminder of how basic technologies of distancing—a wall, a mirror, a checkpoint, a gate—can operate effectively even at close physical range.[4]

And yet under the slow, constant scrutiny of everyday experience, the sense made on casual visitors of the wall's absolute limit yields to an understanding of its porosity, its points of penetration. Take, for instance, the galvanized-steel door at the end of the hallway framed by the office of the (white male) vice president and the glass conference room. Capable of being locked only from the front-office side of the wall, this door offers the only passageway between the front office and the rest of the slaughterhouse. It is the door through which certain documents from the various interior chambers of the slaughterhouse pass in order to be filed for inspection by those who will visit only the front office. In the opposite direction, it is the door through which selected visitors are ushered in order to offer them a sense of initiation, of insider knowledge about what exists behind the opaque southern wall.

Or perhaps more emblematically, consider the wall's one other point of entrance: the small window in the conference room with shades that can be controlled only from the front-office side. One imagines PowerPoint–supported sales presentations culminating in a dramatic pulling back of these shades as prospective buyers swivel in their leather chairs and gaze with frank curiosity through the portal of their wall at what lies beyond.

The sporadic porosity of the wall produces a contradiction inherent in the slaughterhouse. Do not look, you cannot look, you must not look, you do not have to look: this is what the wall says. Look, you can look, it is possible to look, it is necessary to look: this is what the galvanized-steel door and the window say. This call and response between an iron-clad wall and its door and window creates a modern front office that is both attached to and severed physically from the work of killing it enables and feeds upon. It is this same call and response of do not know, you do not have to know and yes know, why would you not know? that characterizes a central dynamic in all contemporary processes of mediation and distancing.

Fabrication

fabricate, v.
1. To make anything that requires skill;
to construct, to manufacture
2. In bad sense: To "make up"; to frame or invent
(a legend, lie, etc.); to forge (a document).
—Oxford English Dictionary

The front office's small square window and galvanized-steel door allow visual and physical movement into the next compartment of the industrialized slaughterhouse: its fabrication department. Here hundreds of handheld knives and saws reinvent chilled half-carcasses as steaks, rounds, and roasts that are then boxed and shipped to distributors and retailers around the world. The work requires manufacturing skill: it is a kind of construction, a kind of craftsmanship. But it is also a work of making up, of framing and invention, an alchemy of deception that authorizes mythical tales—lies, really—about "meat" in contemporary industrialized societies. Cattle half-carcasses enter the fabrication department from the cooler without head, hide, hoofs, or internal organs, but the basic contours of a once-living creature are plain: there the legs, severed just above the hoofs; there the neck and broad shoulders, eerily headless; there the ribs, the back, the bisected spinal cord. By the time these half-carcasses have made their circuit through the fabrication department, they will have been manufactured and constructed into "primal" and "subprimal" cuts, shrink-wrapped, and boxed in ways that render them unintelligible as animal, as a once-living creature. Here there occurs both linguistic and material manufacturing: the fabrication department is a site of production, a hidden workshop floor where the linguistic leap from *steer* to *steak,* from *heifer* to *hamburger* is enacted.

The transition from front office to fabrication department via the galvanized-steel door is severe and startling. Glass and steel give way to walls made of waist-high unpainted concrete at the base and off-white corrugated metal that juts several stories upward before becoming invisible in the harsh glare of evenly spaced circular halogen lights three feet in diameter, suspended from the ceiling by metal rods. The floor is rough concrete: hard, damp, and slippery with water and chunks of fat. The ceiling, visible only with a strong flashlight, is textured with concrete crossbeams as large as the pillars supporting it. There is not a single entry point for natural light, and if the room were cleared of its tables, cutting equipment, air-conditioning fans, hoses, rails, lights, and networks of conveyor belts, only an airless hangar and square concrete pillars would remain. To minimize bacterial growth, an automated cooling system keeps the temperature in the fabrication department near 50 degrees Fahrenheit.

The department is divided into six parallel "tables," wide plastic conveyor belts framed on each side by stationary flat metal working surfaces. Approximately fifty workers and one foreman, all assigned to a particular set of meat cuts, stand shoulder to shoulder on raised webbed platforms flanking each table. Some workers use hydraulic and air-powered cutting instruments; most work with handheld knives ranging from a few inches to more than a foot in length. The dominant hand holds the knife, the other a metal talon anchored in orange plastic to hook and pull chunks of meat from the moving conveyors.

Workers wear long white coats called frocks over sweaters, jackets, long underwear, pants, and ankle-high, steel-toed leather boots. To protect the hands and arms from knife cuts, white cotton gloves are pulled over rubber gloves and wire-mesh forearm sleeves. A white hard hat completes the outfit. Garbed like lab scientists on the verge of invention, the white-frocked fabrication-department workers transform, one cut at a time, carcass into meat. The coats communicate hygiene, cleanliness, and control; they also convey the methodical precision with which the work of fabrication at both its physical and linguistic levels is carried out.

Wherever distance and concealment are at work as mechanisms of power, might we not expect to find analogous fabrication departments staffed by white-frocked workers who undertake the physical task of rendering the repugnant palatable? And might we not also expect to find in these departments a parallel alchemy of linguistic distancing, where things might enter as plainly as bellowing cattle and exit as inert and neutralized as packaged beef? In all political processes where the unacceptable must be rendered acceptable, where the morally and physically disgusting must be made digestible, fabrication departments—literal and allegorical—perform a dual work of construction and manufacture and of framing, forgery, and the invention of legends and lies. Innocents murdered in war become "collateral damage"; the condemned are "executed"; countries are "pacified" and their native populations "dispersed" in "mopping-up operations"; "kinetic operations" "neutralize" and "liquidate" their "targets"—the varieties of fabrication are limited only by the bounds of language and imagination.[5]

The shaded window from the front office to the fabrication department suggests a paradoxical relation between society at large and its acts of fabrication, both physical and linguistic. These acts demonstrate a mastery over perception and are a source of showmanship and pride, but they retain their efficacy only to the extent that the inner workings required to produce them remain out of sight. They are enabling fictions—the words we use in the stories we tell to make the status quo livable. At their sites of production there is both pride and wariness in revealing how the fictions are written.

The Cooler

Someone magically transported to the cooler's entryway would encounter this sight: steaming swaying white half-carcasses and dark red livers flowing in parallel lines along two rails down a wide concrete staircase that has no apparent opening at the top; rows of motionless carcasses interspersed with disembodied tongues and tails stuck two to a hook; and a line of linked carts heavy with more livers, blood trickling slowly out of the puncture wounds where the hook has been forced through them. The cooler is an unsettling land of in-between where bodies and body parts, neither whole nor completely disassembled, are recognizable as individual entities—this tail, this carcass, that tongue, that liver—but arranged in rows and lines of sufficient mass that the mind struggles to imagine the sheer scale of the overall puzzle of which they are the pieces.

The cooler is a series of large rooms that lie like a sparsely populated tundra separating the humid jungle of the kill floor from the controlled manufacturing environment of fabrication. As in the fabrication room, walls of waist-high concrete are topped by off-white corrugated metal several stories in height. Massive square concrete pillars sit at regular intervals, girded by horizontal concrete beams that span the ceiling's width. There are no windows, no outside light or air. Large halogen lamps hang from slim metal rods, spaced far enough apart that bright spots of light shade imperceptibly into gray darkness before intensifying again under the power of a different lamp.

But unlike the fabrication room, where space is organized from below, by the tables, the space of the cooler is divided from above: long, horizontal metal rails hang suspended from the ceiling and stretch in parallel lines across the width of the cooler's chambers. These rails connect on each end to two rails that run the length of the cooler, connecting its chambers, and lead into the fabrication department. Every few feet along the rails are metal spacers that under pressure fold flat against the rail, forcing the half-carcasses to remain evenly spaced on their one-way journey through the cooler. The two "length" rails are the main throughway tracks: half-carcasses enter the cooler from the kill floor on one length rail and leave the cooler twenty-four to forty-eight hours later on the other. During their time in the cooler, the half-carcasses hang, completely still, on one of the hundreds of "width" rails, forty-eight half-carcasses per rail. Each morning, starting at the chambers closest to the fabrication department, chilled carcasses are moved off the width rails onto the length rails leading into fabrication. Freshly killed carcasses arriving on the other length rail from the kill floor are then railed onto these vacated width rails until each chamber fills to capacity.

The entrance to the cooler lies at the bottom of a staircase called the decline, a set of concrete steps about thirty feet wide set between cinder-block walls. A corrugated-metal wall frames the top of the decline; the opening to the kill floor sits at a right angle to this wall, hiding the work of the kill floor from that of the cooler. The overhead rail transporting the split carcasses exits the kill floor and makes a 90-degree turn via a sprocket down the center of the decline. Suspended from this overhead rail, half-carcasses float from the kill floor through the opening at the top of the decline and descend to the cooler's entrance, where a heavy, insulated, fifteen foot–high door is chained open. To keep the cooler insulated, this door is closed during breaks in production; because it opens outward against the flow of oncoming carcasses, the door must be opened before the first carcass of the day or the first carcass after a work break reaches the entrance to the cooler.

At the start of each day, the decline is dry and spotless. By day's end, the concrete steps are visible only in patches beneath globs of fat and pools of blood shaken free by the jolting overhead chain as it lowers the four hundred–pound half-sides down the decline. Throughout the workday the decline is steamy, hot, and loud with the clanking of the overhead rail and the thud of carcasses swinging into one another. Walking up or down the decline under these conditions is risky, as is moving between the swinging carcasses to get from one side of the decline to the other.

Inside the cooler, the temperature is just above freezing, between 33 and 34 degrees. Spaced evenly along the width rails, overhead sprinklers turn on and off automatically, spraying water that speeds up the chilling of the carcasses. Long thin drains covered by metal grates line the cooler floor, one drain between each pair of width rails. Just under the ceiling, enormous fans hum deafeningly, low-altitude jet engines trapped perpetually in place without possibility of flight. Smell in the cooler shades from hot, pungent, and ironlike where the decline meets the cooler entranceway to sterilized, crisp, and refrigerated one or two chambers into the cooler.

Half a dozen workers called railers keep the waves of half-carcasses flowing smoothly in and out of the cooler. They wear multiple layers of warm clothing under bright yellow rain suits and white hard hats. White cotton gloves cover green waterproof rubber ones. Their main implements are two short metal hooks with orange plastic grips, which they sink into the half-carcasses to stop, push, or pull them manually along the rails. Control boxes mounted on the wall of the cooler allow the railers to operate hydraulic switches that control which width rail connects to one of the two length rails. At the cooler's entrance, two railers stand like sentries, one on either side of the doorway, using a sponge mounted at the tip of a fifteen-foot metal pole to wipe down the potentially contaminating condensation that forms on the overhead rails when the rush of hot, humid air from the kill floor collides with the cooler's frigid atmosphere. Craning their necks to spot the condensation, these railers use their long poles to dab at the overhead rails in the miniscule intervals of space and time between the carcasses entering the cooler from the decline. Their second job is to watch for specially tagged carcasses that must be railed near the front of the cooler onto special rails designated for samples selected for random bacterial testing or for carcasses of older cattle at increased risk of harboring bovine spongiform encephalopathy (BSE). When a specially tagged carcass approaches the cooler entrance, the railers must act quickly, activating the switch connecting the special width rail to the incoming length rail so that the marked carcass is directed onto it, then deactivating the switch again before the next unmarked carcass whizzes by less than twelve seconds later.

In addition to the yellow-jacketed railers posted throughout the cooler and at its entrance, two other categories of workers occupy the cooler. One group works in the area just before the fabrication department and is auxiliary to that department. These workers wield handheld air-compressed knives and make horizontal incisions in the carcasses at a midpoint in the ribs in order to prepare them for disassembly in the fabrication department. The other group is located near where the decline enters the cooler and is auxiliary to the kill floor. This group's job is to handle the different body parts—tongues, tails, and livers—that require extended chilling before packing and shipment.

The cooler's punishing cold yields an unexpected benefit: a relative dearth of supervision. Both kill floor and fabrication department auxiliary cooler workers have foremen, as do the yellow-jacketed railers, but these supervisors are also needed in other areas of the slaughterhouse, making it impractical for them to wear the many layers of warm clothing necessary for staying in the cooler for an extended period of time. The result is a relative hinterland of comparative freedom, especially for the railers, whose work gives them license to roam the vast space of the cooler. Anarchists of the slaughterhouse, these yellow-jacketed workers find hiding places between thick, silent rows of carcasses and the uneven spotlights cast by the overhead lamps. In these lightly patrolled territories, long metal poles for wiping condensation off overhead rails become lances for sword fights, duels, and poking unsuspecting coworkers in the back. Pieces of fat, shaken loose from the carcasses and rapidly hardening on the cooler floor, become projectiles for launching. Massive concrete pillars become leaning posts for unauthorized breaks or hiding places from cold-harried supervisors and inspectors passing rapidly by. Corrugated-metal walls and pools of blood become whiteboard and dry-erase ink, suitable for drawing cartoons lampooning unpopular supervisors or for an impromptu English lesson. The frigid tundra shielding fabrication from kill floor is also a place of spontaneity and play, belying what lies just beyond.

III
Kill Floor

Job Number 3, Knocker*: operates knocking box; uses air gun to drive captive-steel bolt into foreheads of cattle while they are suspended on center track and restrained by sidewalls.*

No direct route connects the kill floor and front office. The quickest way to move from one to the other is to leave the building and walk around the perimeter. Otherwise, travel between the two necessitates a circuitous route through the cooler and the fabrication department. The kill floor and front office are as far apart physically as possible without being separated into two distinct buildings, an isolation that is mirrored bureaucratically. Whereas front-office staff supervise the fabrication department, the kill floor administrators are housed in offices on the floor. In this state of physical and bureaucratic isolation from both society at large and the rest of the slaughterhouse, the kill floor throws off the euphemistic linguistic blanket cloaking the slaughterhouse's other departments. *Kill* floor: there is a brutal honesty to the name, an attempt to create distance between the insiders who take lives and the outsiders, both societal and within the slaughterhouse itself, who directly rely on, but are not privy to, that taking.

In an era in which words are sanitized, packaged, and re-presented in much the same way that meat is cellophane wrapped in featureless, uniform packages, *kill* floor is the linguistic equivalent to showing up at a dinner party with a squawking chicken and wringing its neck in a living room full of wine-sipping guests before going into the kitchen and plopping it into the pot. Indeed, meat-industry and animal-science textbooks train graduates to refer to killing as "harvesting" and to the kill floor itself as a "harvesting department." This Orwellian attempt at re-education does not seem to have succeeded, however, and the dysphemistic *kill floor* remains the official and unofficial name of the place dedicated to the taking of life; its workers are called and call themselves, without pretense, *kill floor workers.*[1]

These nomenclature differences, tricks of language that locate the kill floor a linguistic universe away from the fabrication department, also extend to the kill floor's working environment, dress codes, and supervisory and inspection regimes. In fabrication, the air is kept at a near constant temperature, while the chilling action of the cooler congeals the blood in the animal's muscles. All the animal's internal organs have been removed. Dulled by the cold, smells are almost nonexistent. The cooler and fabrication departments work with solids, which makes them more sanitary than the kill floor, where leaking fluids—from blood to urine to feces to vomit to bits of brain matter to bile—are a constant presence: on the floor, on the walls, on the machinery, and on the knives, clothes, and bodies of the workers themselves. In contrast to the crisp, cold air of fabrication, the air of the kill floor is steamy and humid; each time a carcass is split open, it emits more heat and humidity into the room. The smell on the kill floor varies from place to place, but it is always fiercely organic, a combination of feces, urine, vomit, brain matter, and blood in various stages, from fresh to congealed.

The homogenization of the animal that takes place in the fabrication department corresponds to a more disciplined, bureaucratized, and predictable production regime. The living creature, the *animal* that is herded off a truck and into the production sequence of the kill floor, in contrast, arrives in varied shapes and sizes, each distinct, each unique. Some balk when prodded up the chute leading to the kill box, some collapse from exhaustion or disease, some have horns that are especially difficult to cut off, some are pregnant and about to give birth, some are unusually large, and some are unexpectedly small. The kill floor must make concessions to this uniqueness, this regular irregularity. In tandem with the cooler, its function is to erase individuality and produce in its place a raw material, an input. Already stripped of all individuating characteristics of hide, horns, and sex, the carcass that reaches the cooler is further homogenized: the very texture of the flesh is reduced to one temperature, one consistency, one thing identical to the thing next to it, which is identical to the thousands of things next to it, all ready to be fabricated into a series of meat "products."

In fabrication, workers wear identical white frocks and stand side by side, giving the appearance of a solid mass of interchangeable units. Compared to the fabrication department, the kill floor seems sparsely populated. Kill floor workers stand at least one to two feet apart. Although the floor space is much larger than that of the fabrication department, the kill floor employs only slightly more than half the total number of workers found in the fabrication department. Workers on the kill floor have more personal space, more room for variation in their bodily motions (although these, too, are highly constricted by the fast pace of the line). The difference in density is due, in part, to the fact that the arrangement of production space on the kill floor is dictated by the ordering necessitated by killing, dehiding, and eviscerating a whole animal. Killing must precede dehiding, which must precede evisceration, and so forth. Simultaneous parallel production lines are possible only after the animal has been killed, decapitated, and eviscerated.

Kill floor workers wear their own clothing, typically T-shirts, ranging from plain solid colors to those bearing messages such as "Another life changed by Jesus Christ" to those with picture prints of the fantasized good life: a luxury convertible with a beautiful couple kissing in the back seat or a scantily dressed woman with the words "My Dream Woman" below her. In fabrication, the atmosphere seems somber, almost sedated. Despite their close proximity to one another, workers toil quietly, without conversation or eye contact. On the kill floor, the atmosphere is more boisterous. Songs, shouts, and whistles often pierce the air, and ongoing conversations between workers stationed together on the line are common. Behind the backs of supervisors, workers throw bits of fat around and shoot rubber bands at one another.

This difference in working culture parallels a difference in supervisory control. In fabrication, the parallel arrangement of the tables and the close proximity of the workers make surveillance not only easier but also, from the workers' point of view, less predictable. From any raised point in the fabrication plant, it is possible to see almost every worker, and workers can be watched without their being aware of it. On the kill floor, the linear flow of production makes surveillance both more difficult and more predictable. A supervisor must walk down the line to see what is happening at the other end of the kill floor. Workers communicate by whistles, facial expressions, and hand signals when a supervisor or inspector is on the way. In addition, there is a marked difference between fabrication and the kill floor in the USDA inspection regime. In fabrication, the uniformity and regularity of the production process allows one federal meat inspector to oversee the entire department. On the kill floor, the individuality and irregularity of the animals necessitates the continual presence of at least nine inspectors.

Unlike the single-story operations in fabrication and the cooler, the kill floor takes up two floors. Central functions related to the cattle, carcasses, and offal take place on the upper floor, and supporting operations, such as the maintenance shop, hook cleaning, chemical storage, box making, the boiler room, the air-compression room, and the pet-food room, are on the first floor.

Figure 2, "Overview of the Kill Floor," maps the kill floor of the industrialized slaughterhouse, while figures 3–7 provide details of various areas of the overview. Originally hand drawn in painstaking fragments over many months of fieldwork, the map offers a highly comprehensive overview of the division of labor and space on a contemporary industrialized kill floor. The specific location of each worker is represented by a circle. The numbers inside each circle, from 1 to 121, stand for particular job functions on the kill floor. In some cases, more than one circle has the same number, indicating that more than one worker performs that particular task. The kill floor jobs are ordered consecutively, starting with the unloading and weighing of live cattle and proceeding along the main carcass line to the cooler, where the split and eviscerated half-sides are chilled before continuing to the fabrication department. Auxiliary operations in the gut and foot rooms and supporting sanitation workers are marked 104 through 121. The dark line running through the map indicates the overhead rail that carries the suspended cattle after they have been stunned and killed. In addition, as indicated in the map key, other lines carry various body parts as they are eviscerated or severed from the main body of the animal. Supervisors, quality-control workers, and USDA inspectors are marked with special patterns indicated in the key. Appendix A, "Division of Labor on the Kill Floor," provides a narrative description of each of the 121 job functions on the kill floor. Like the figure, these descriptions are drawn from several months of personal observation of the performances of the tasks.

You can reach a profound understanding of the division of labor and space in contemporary industrialized killing by studying each of these figures in conjunction with the job descriptions provided in Appendix A. Trace the path taken by the cattle on the kill floor map starting with job 1, and actively imagine the actions performed on them and how these transform the individual cow, steer, or heifer as it moves through the kill floor. At what point is the animal killed? Where does it lose its tail? Its hoofs? Its hide? Its head? Its heart, lung, liver, and intestines? Consider also what each individual worker is able to see, from his or her particular vantage point on the kill floor. What does the animal look like to the individual worker

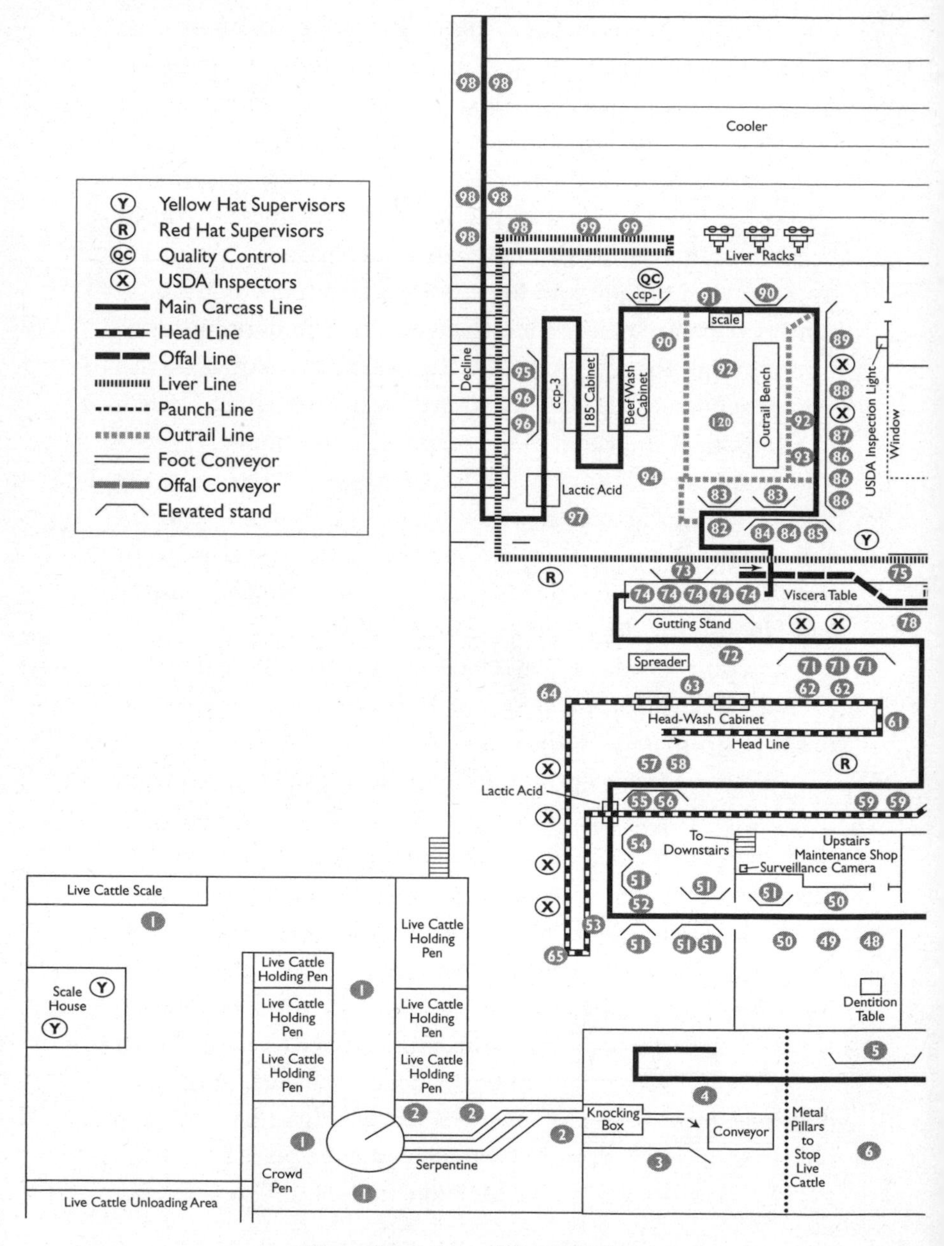

Fig. 2 Overview of the Kill Floor

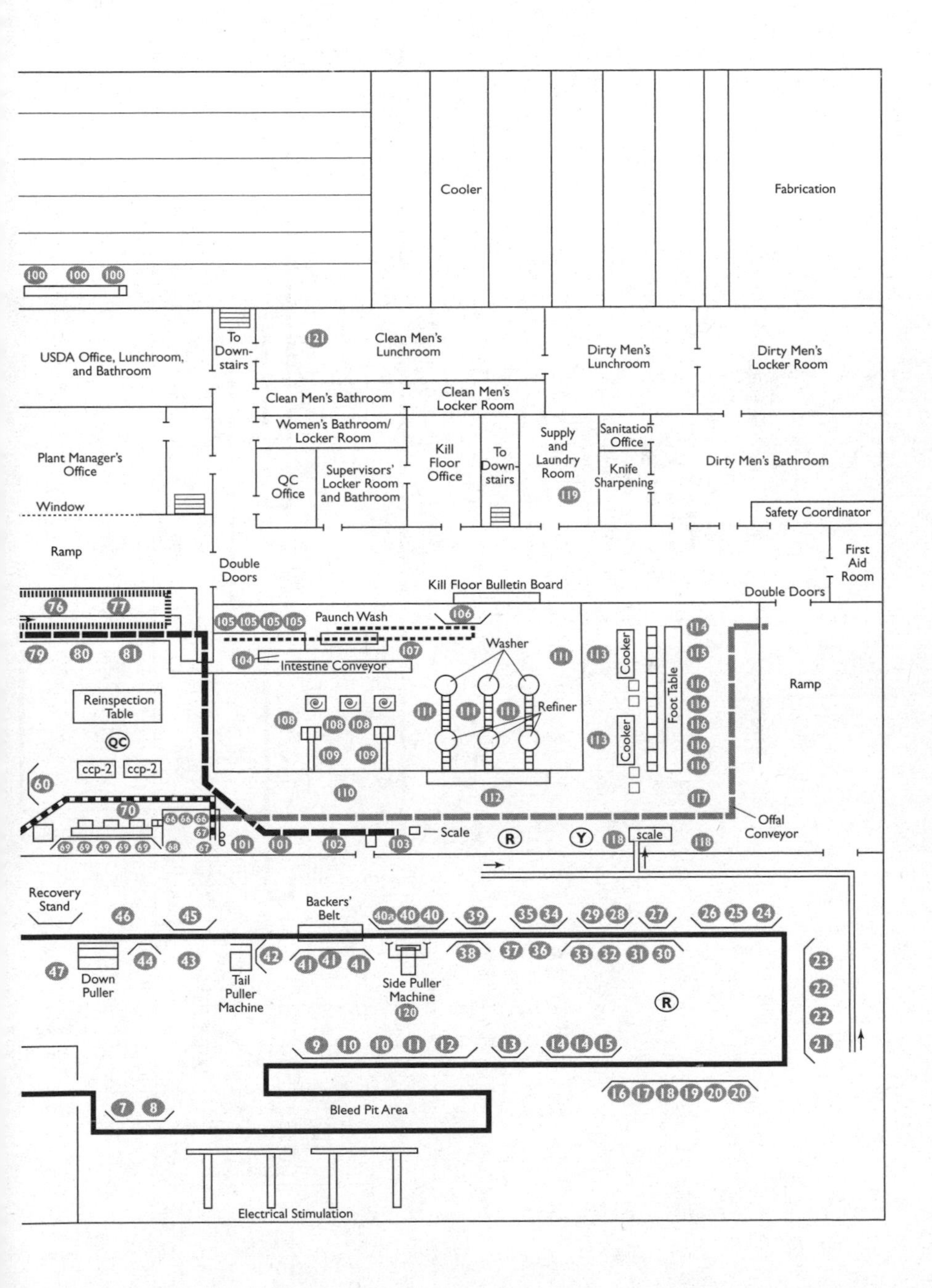

Cooler
Fabrication
USDA Office, Lunchroom, and Bathroom
To Down-stairs
Clean Men's Lunchroom
Dirty Men's Lunchroom
Dirty Men's Locker Room
Clean Men's Bathroom
Clean Men's Locker Room
Women's Bathroom/ Locker Room
Plant Manager's Office
QC Office
Supervisors' Locker Room and Bathroom
Kill Floor Office
To Down-stairs
Supply and Laundry Room
Sanitation Office
Knife Sharpening
Dirty Men's Bathroom
Window
Safety Coordinator
Ramp
Double Doors
First Aid Room
Kill Floor Bulletin Board
Double Doors
Paunch Wash
Intestine Conveyor
Washer
Refiner
Cooker
Cooker
Foot Table
Ramp
Reinspection Table
QC
ccp-2
ccp-2
Scale
scale
Offal Conveyor
Recovery Stand
Backers' Belt
Down Puller
Tail Puller Machine
Side Puller Machine
Bleed Pit Area
Electrical Stimulation

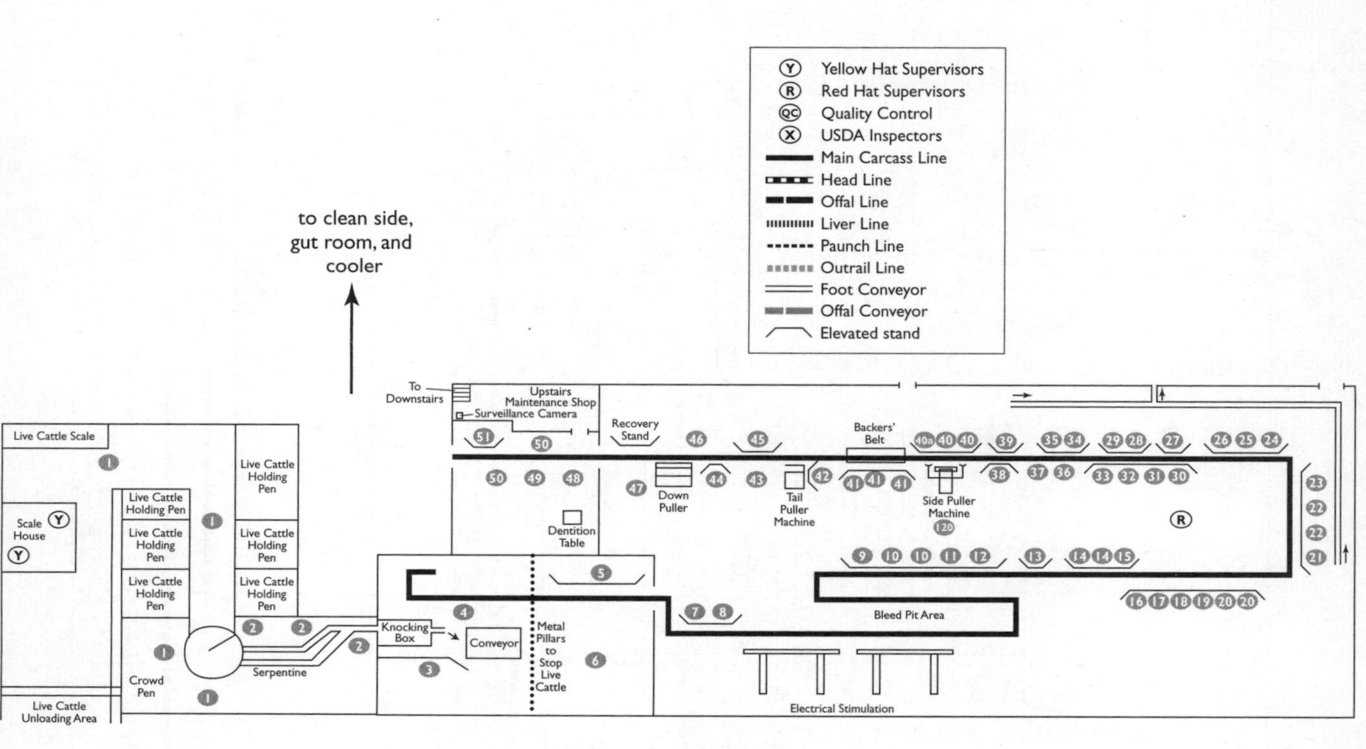

Fig. 3 The Dirty Side

as it passes? The "sticker" represented by circle 8, for example, sees something radically different from what is seen by the "spinal cord removers" represented by the circles numbered 84, and this is yet again completely distinct from what is seen by the "omasum and tripe washers and refiners" represented by the circles numbered 111. There are 121 job functions, 121 perspectives, 121 experiences of industrialized killing.

As a complement to a step-by-step understanding of the linear ordering of killing, the overall space of the kill floor might also be usefully conceptualized as a series of nonexclusive pairs, each describing an important physical, functional, or experiential distinction. These pairs include inside versus outside, alive versus dead, clean versus dirty, main line versus auxiliary, upstairs versus downstairs, and supervisory versus production, and they will help us interpret the divisions of labor and space that organize work on the kill floor.

Inside/Outside

The separation of the cattle-unloading area and pens from the interior of the plant, where the killing and carcass disassembly take place, can be seen as an inside/outside distinction. The cattle-unloading area and pens are not literally outdoors; they are contained in a semi-enclosed structure with head-high concrete walls that support a raised tin roof resting on metal bars. The space between the walls and the roof is open, but the walls block the line of sight from the outside. The floor of the cattle-unloading area is made of brick blocks and is divided into a series of pens walled with tubular metal. Four pen workers (circle 1 on the map) move the cattle in lots from the holding pens to the squeeze pen, a circular enclosure with a large gate capable of being pushed on rollers to "squeeze" or narrow

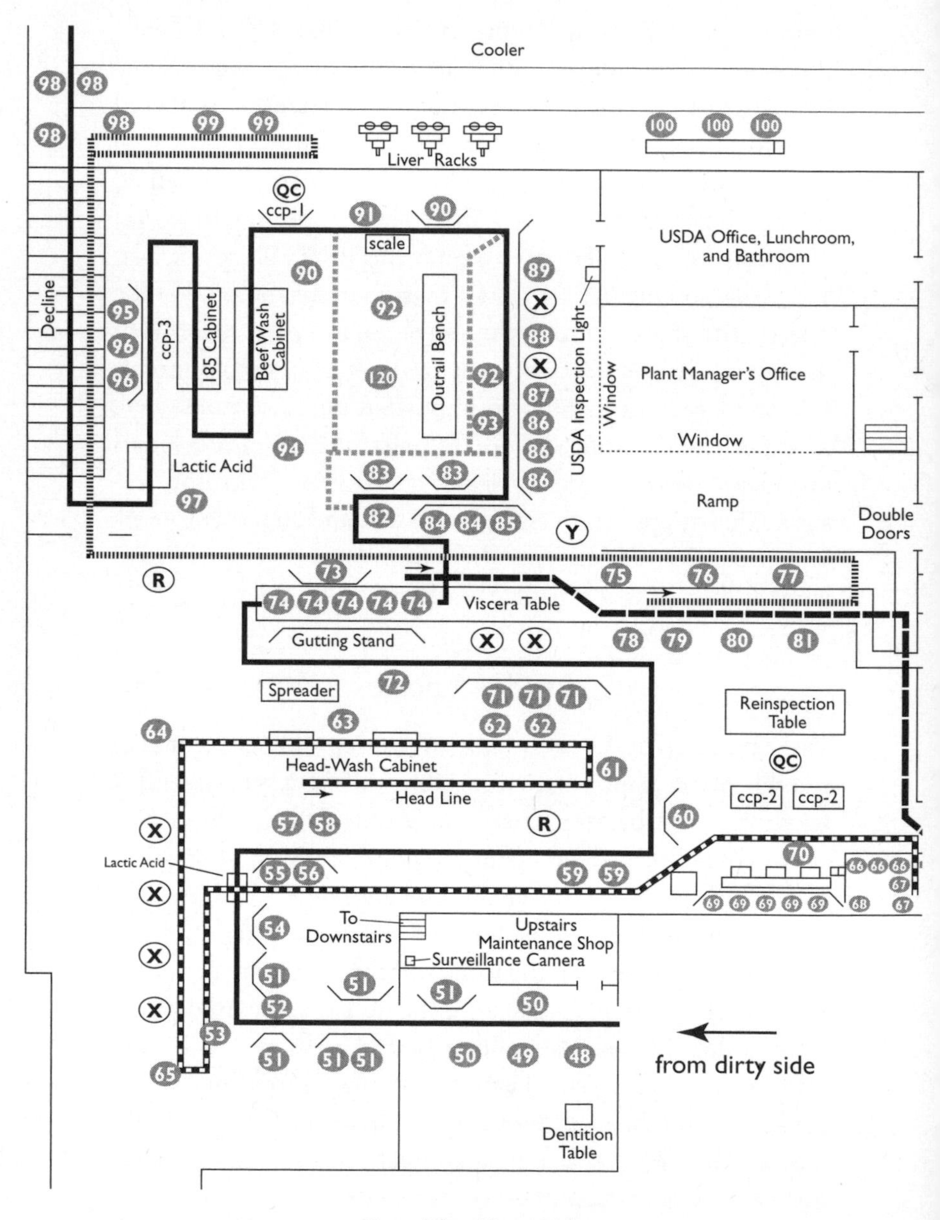

Fig. 4 The Clean Side

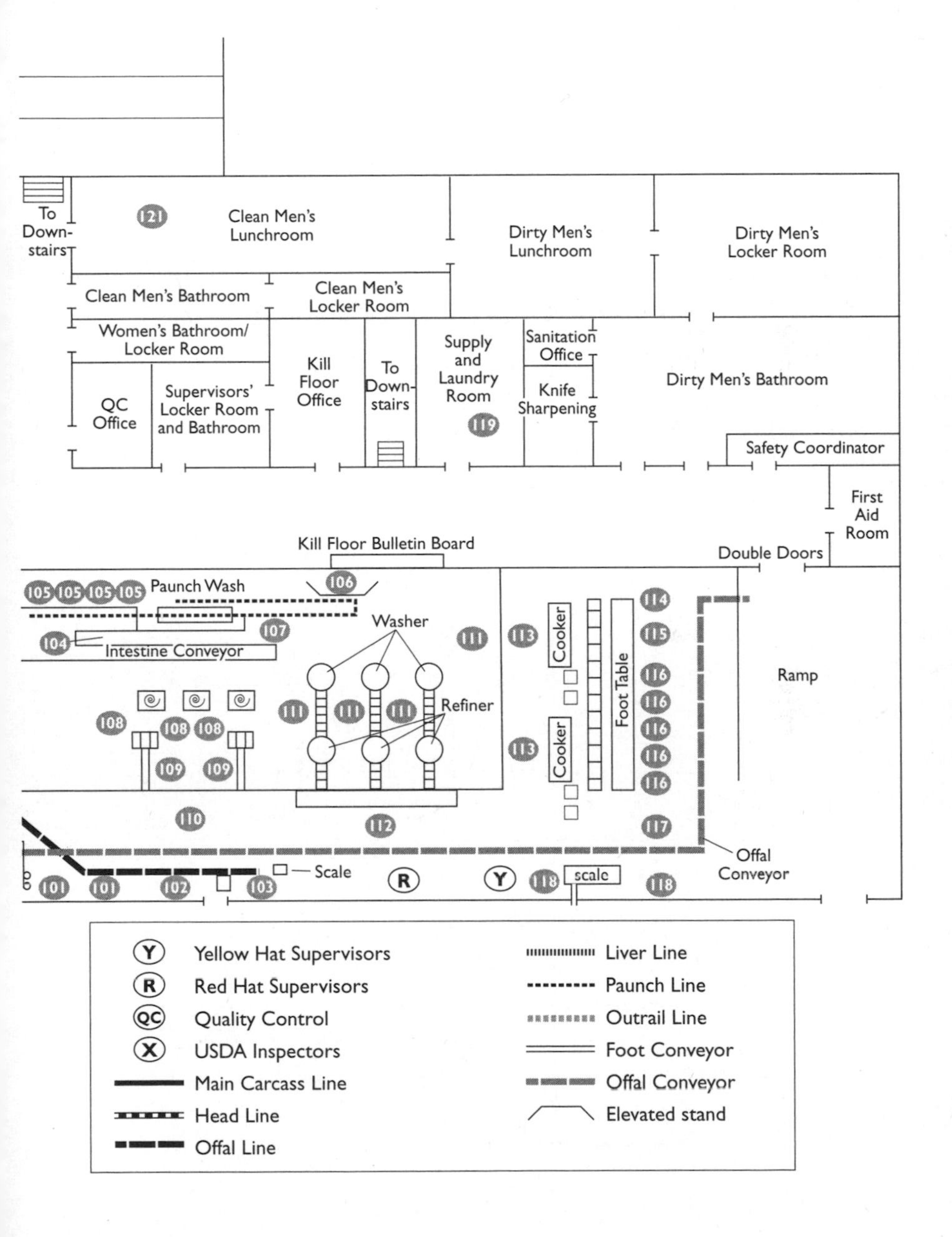

To Down-stairs
121
Clean Men's Lunchroom
Dirty Men's Lunchroom
Dirty Men's Locker Room
Clean Men's Bathroom
Clean Men's Locker Room
Women's Bathroom/ Locker Room
Kill Floor Office
To Down-stairs
Supply and Laundry Room
119
Sanitation Office
Knife Sharpening
Dirty Men's Bathroom
QC Office
Supervisors' Locker Room and Bathroom
Safety Coordinator
First Aid Room
Kill Floor Bulletin Board
Double Doors
105 105 105 105
Paunch Wash
106
104
Intestine Conveyor
107
Washer
111
113
Cooker
Foot Table
114
115
116
108
108 108
109 109
111 111 111
Refiner
Ramp
110
112
117
Offal Conveyor
Scale
scale
101 101 102 103
R
Y
118 118
Yellow Hat Supervisors
Red Hat Supervisors
Quality Control
USDA Inspectors
Main Carcass Line
Head Line
Offal Line
Liver Line
Paunch Line
Outrail Line
Foot Conveyor
Offal Conveyor
Elevated stand

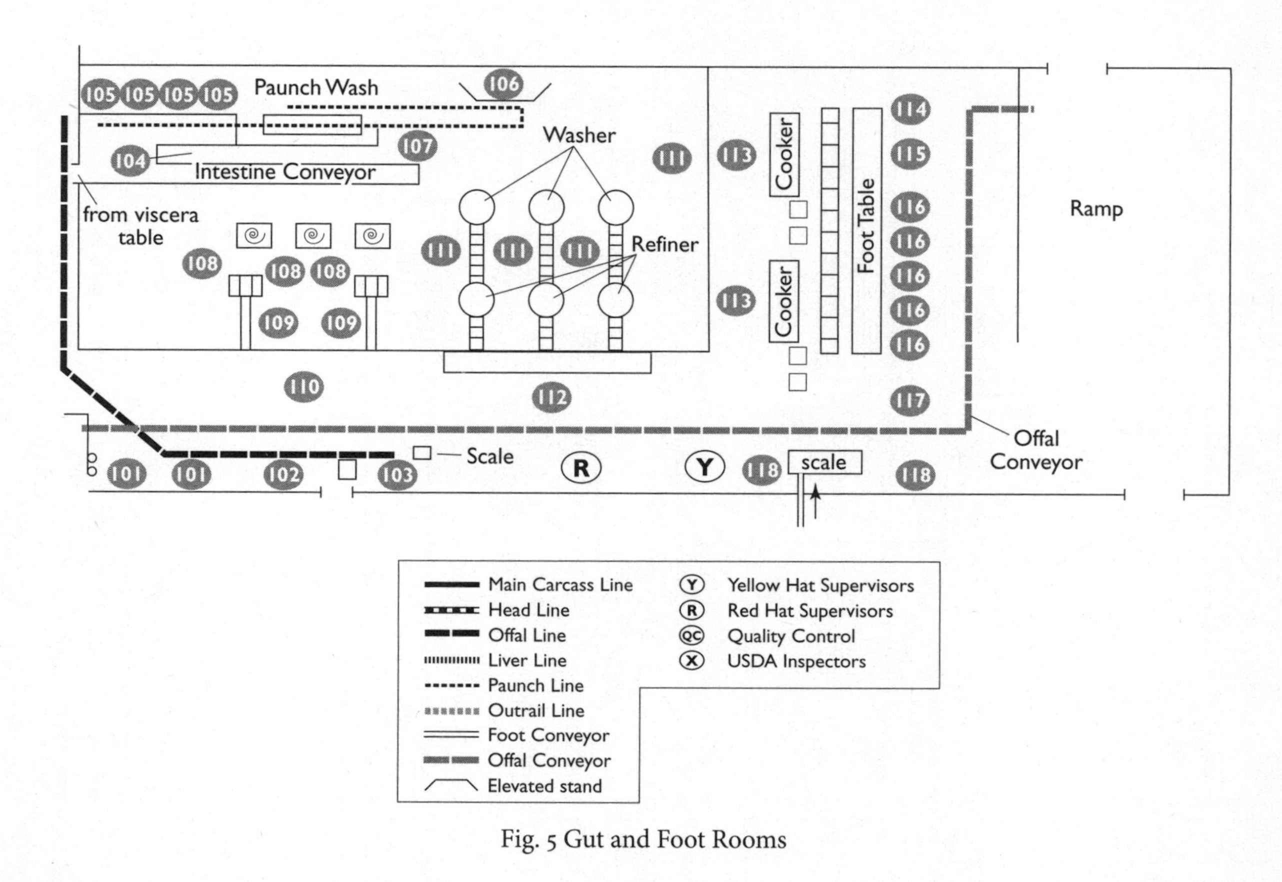

Fig. 5 Gut and Foot Rooms

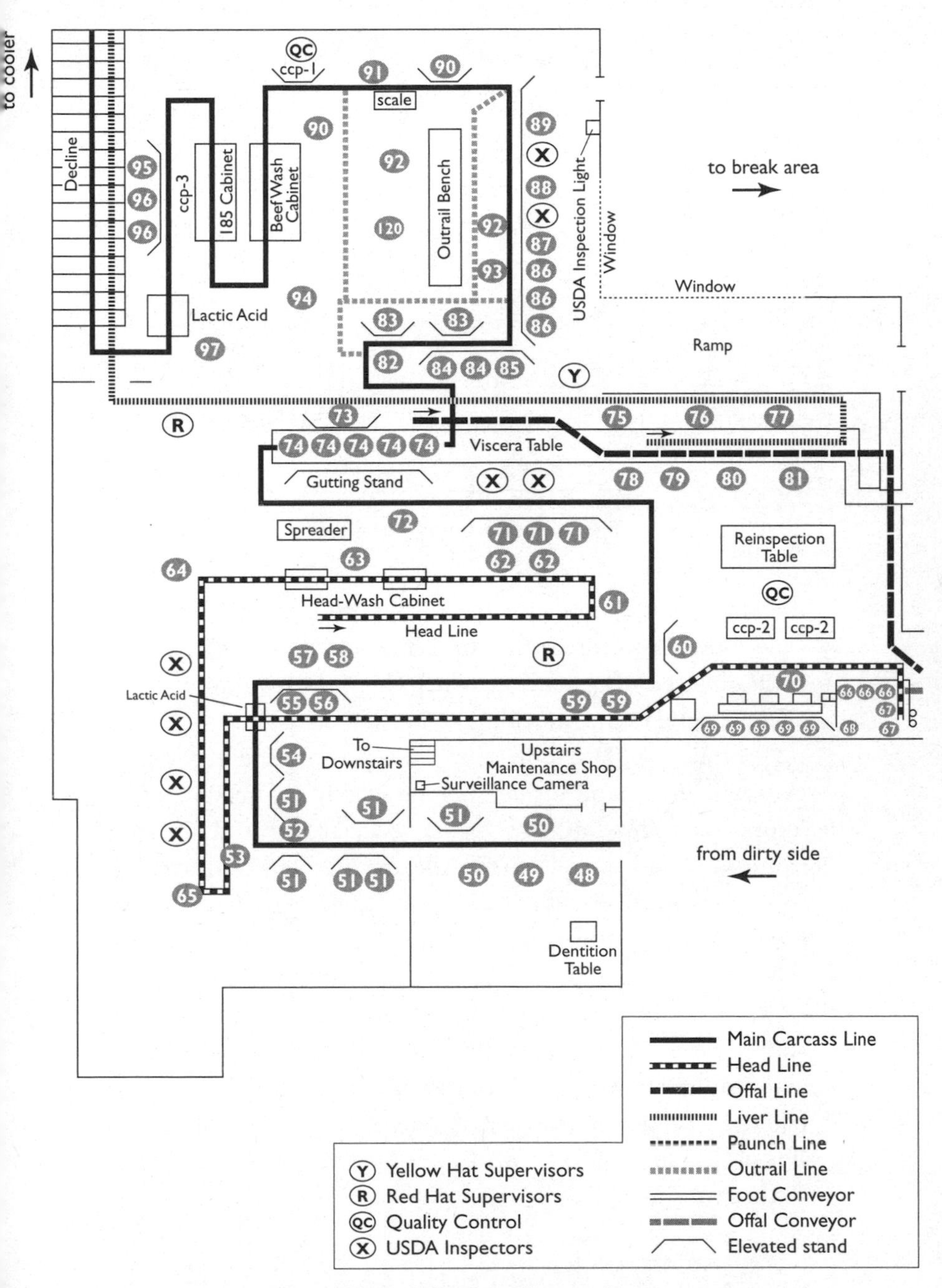

Fig. 6 Clean Side Main Carcass Line

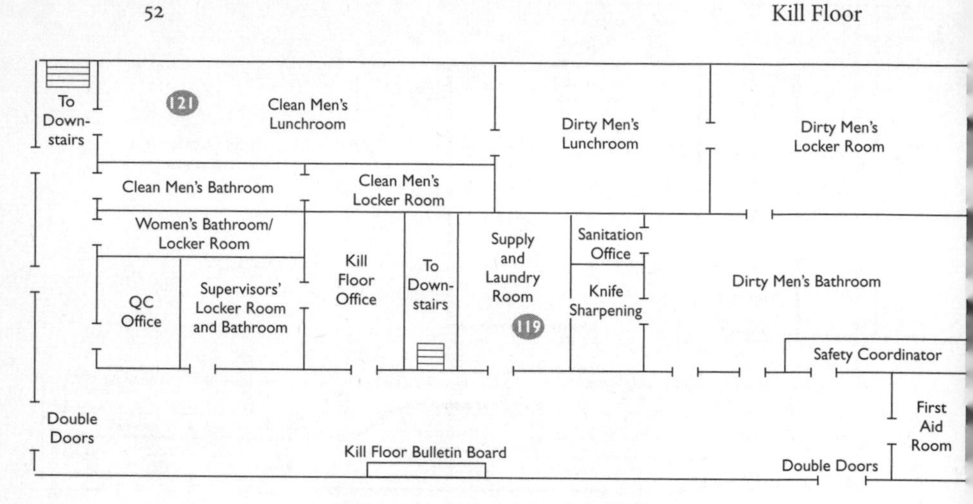

Fig. 7 Break Rooms and Offices

the amount of space available to the cattle in the pen. This squeezing action forces the cattle into one of two serpentine chutes located near circle 2, which then narrow into a single chute leading to the knocking box inside the plant.

From a USDA food-safety perspective, the separation between outside and inside must be strictly enforced to avoid "cross-contamination," the transference of harmful bacteria from the outside area to the inside area. In reality, movement between the two spaces is continuously taking place. First and foremost, live cattle enter the inside of the slaughterhouse through the chutes and a raised trapdoor that separates the chutes from the knocking box (near circle 3). In addition, pen and chute workers constantly walk through the production plant to the toilets, locker rooms, and lunchrooms. Even the USDA inspection officials move between the pens and the inside of the kill floor to perform antemortem inspections of the cattle in the pens.

Alive/Dead

Although the precise point that separates life from death in the slaughterhouse is located somewhere in the electrical stimulation and bleed pit area (near circle 8), the actual killing begins just inside the walls of the slaughterhouse and continues for another fifty or so feet along the line. This killing process occurs in two stages, each stage located out of the direct line of sight of the other. The first stage is the knocking box, the second the presticker (circle 7) and sticker (circle 8).

Cattle are driven through the squeeze chute and up the serpentine. Once the animal passes under the raised trapdoor that separates outside from inside, it is forced onto a metal ramp that angles down into a large dark metal box. At the top of the box the cattle are sprayed with mist from a nozzle. At this point, the ramp gives way to an inverted U-shaped metal conveyor, which supports the cow's underbelly, leaving its legs suspended in the air. The knocker hydraulically controls the movement of this conveyor, as well as the two side walls, which can move inward to constrict the cow. The knocker presses a button that moves the metal conveyor forward, bringing the cow's head out of the rectangular box. If the cow thrashes or struggles, the knocker activates the side walls to further constrict it. Suspended on the conveyor and squeezed by the side walls, the animal can now move only its head.

Once the cow is immobilized, the knocker takes hold of a long, air-powered metallic gun suspended by a counterbalanced cable from an overhead bar. The gun is cylindrical, about a foot long and eight inches in diameter, and it weighs about ten pounds. Seizing what may be only a fraction of a second when the cow's head is still, the knocker angles the gun down and presses it against the cow's head, between and slightly above the eyes. The pressure of the gun's snout against the cow's head releases the safety, much as an air-powered nail-gun operates. The knocker pulls the trigger, which releases a retracting cylindrical steel bolt approximately five inches long and an inch in diameter. The bolt penetrates the cow's skull, then quickly retracts. The sound made by the firing of the bolt and its impact is a muted *pffft, pffft.* As the bolt retracts, gray brain matter often flies out of the hole in the cow's skull, sometimes splattering the clothing, arms, or face of the knocker. Seconds later, blood gushes out of the wound, bubbling up and out in a dark maroon stream as it oxygenates. Sometimes the cow's head will immediately drop, hitting the metal conveyor or (if the cattle are spaced closely together) falling onto the rump of the cow in front of it. At other times the cow's neck will stiffen, with its head locked unnaturally face up in the air. When this happens, the neck and head tremble at a high speed, as if in a seizure. Whether the head falls or the neck stiffens, the cow's eyes typically take on a glazed look, and its tongue often hangs limply from its mouth. Sometimes the power, angle, or location of the steel bolt shot is insufficient to render the cow unconscious, and it will bleed profusely and thrash about wildly while the knocker tries to shoot it again.

After the cow has been shot, the knocker advances the conveyor, and the cow drops onto another conveyor, of wide green plastic, about five feet under the metal conveyor. Because the cow is unconscious at this point, it often falls forward onto its head, sometimes breaking its teeth or biting its tongue. Once the animal is on the plastic conveyor, the shackler (circle 4) wraps a metal hook around its left hind leg. The hook is suspended from a chain connected via a wheel to an overhead rail. The rail moves the wheel forward, lifting the cow into the air by its left hind leg until it is suspended vertically, head down. The cow's right hind leg and front legs often begin to kick wildly at this point, creating the impression that the cow is still alive and conscious. Meat-industry publications state that these motions are purely reflexive and do not indicate consciousness; the key to establishing consciousness, they claim, lies in the tongue and the eyes. If it has not done so already, the cow will often vomit, depositing a rank greenish substance onto the floor that mixes with the blood flowing from its head wounds.

At this point, there is no spacing mechanism on the overhead rail to separate the cattle so that they hang at equal distances. The result is that they are bunched tightly together, producing at the top a synchronized row of paired legs, one immobilized by the chain and the other kicking wildly, and at the bottom a drizzling red and green liquid screen of blood and vomit falling onto the slaughterhouse floor. The indexer (circle 5), standing on a raised stand, begins spacing the cows with a long metal pole at regular intervals, inserting a metal link, called a dog, between each cow on the overhead rail. One cow hangs between each pair of evenly spaced dogs, creating a predictable rhythm of work for all the subsequent workers on the line. The indexer also watches for any signs of consciousness among the cattle that have just been shot. These include attempts by the cow to right itself, reflexive blinking in response to stimuli, and a tongue that is not hanging limply from the mouth. If the indexer notices any of these, he takes a captive-bolt handgun powered by a bronze cap that looks like a .22 shell and fires into the head of the cow. Unlike the soft *pffft, pffft* of the knocker's gun, this handgun makes a sharp report not unlike the crack of a rifle and releases the acrid smell of gunpowder into the air.

Just past the indexer, the overhead rail makes two sharp 90-degree turns. The first takes the cattle behind a wall that isolates the knocking box and indexer stand from the rest of the kill floor; the second reestablishes the original direction of the overhead rail. The presticker (circle 7) and sticker (circle 8) are located directly after this second 90-degree turn, shielded by the wall from the line of sight of the knocking box and indexer. The presticker and sticker work together on another raised stand. As the cattle come by, the presticker takes a knife and makes a vertical incision in the neck of each cow. Because the cows are still kicking reflexively, he must approach them from the side to reduce his exposure to their legs. The hand that is not gripping the knife is often held up in front of his face as the presticker lunges forward, makes the incision, and retreats to a safe distance as quickly as possible.

The sticker stands about three feet from the presticker. Using a hand knife, he reaches into the vertical incision made by the presticker to sever the cow's carotid arteries and jugular veins. This cut produces a gushing torrent of blood that flows through the webbed platform onto the floor. An accomplished sticker can judge precisely the direction and volume of this flow and deftly sidestep it to avoid being drenched by the blood.

Technically, it is the severing of the carotid arteries and jugular veins that kills the cow, which will die somewhere in the electrical stimulation and bleed pit area located immediately after this workstation. The electrical stimulation and bleed pit area is created by the doubling back of the overhead rail on itself in an S shape. A metal-lined trench below the rail catches most of the blood and drains it into a tank on the floor to be stored for subsequent sale (see Appendix B for the various uses of the cattle body parts). The stimulators consist of two electrified metal cross bars anchored to the side of the slaughterhouse wall. As it passes, the cow makes contact with the bars and is jolted and shaken by the voltage, which stimulates the dying heart, forcing it to pump blood through the body and out the severed veins. By the time the cow reaches the tail ripper (circle 9), it is supposed to be completely lifeless.

As illustrated in figure 8, the space separating life from death on the kill floor is an important one, a fact demonstrated by the problems that arise if the animal either dies too early or lives too long. Cows sometimes collapse in the serpentine chute leading to the kill box because of disease, exhaustion, or the nervous stress of being prodded repeatedly with electrical shockers. If the cow collapses in the chute after the point where the two parallel chutes have narrowed into a single one, the passageway where the live cows enter the slaughterhouse becomes completely blocked, causing a major crisis.

If a cow survives the knocker's bolt and the knocker does not stop the metal conveyor in time, the live cow can fall onto the green-plastic shackling conveyor before it has been stunned. If it does, it usually struggles off the belt and begins to run around the kill floor, panicked by the blood, vomit, and sight and smells of the stunned and shackled cows dangling overhead. This happens often enough that special metal corral gates have been built between the knocking box and the indexer stand (the dotted line on the map) to keep the escaped cow from running out onto other parts of the kill floor. When a cow escapes, the shackler must climb a ladder to the kill box platform to avoid being injured by the panicked cow. Meanwhile, the knocker blows his air horn three times, which brings a supervisor, who calls the plant manager on the radio. The plant manager comes to the garage door behind the shackler with a .22 rifle, opens the door (there is a metal corral gate here

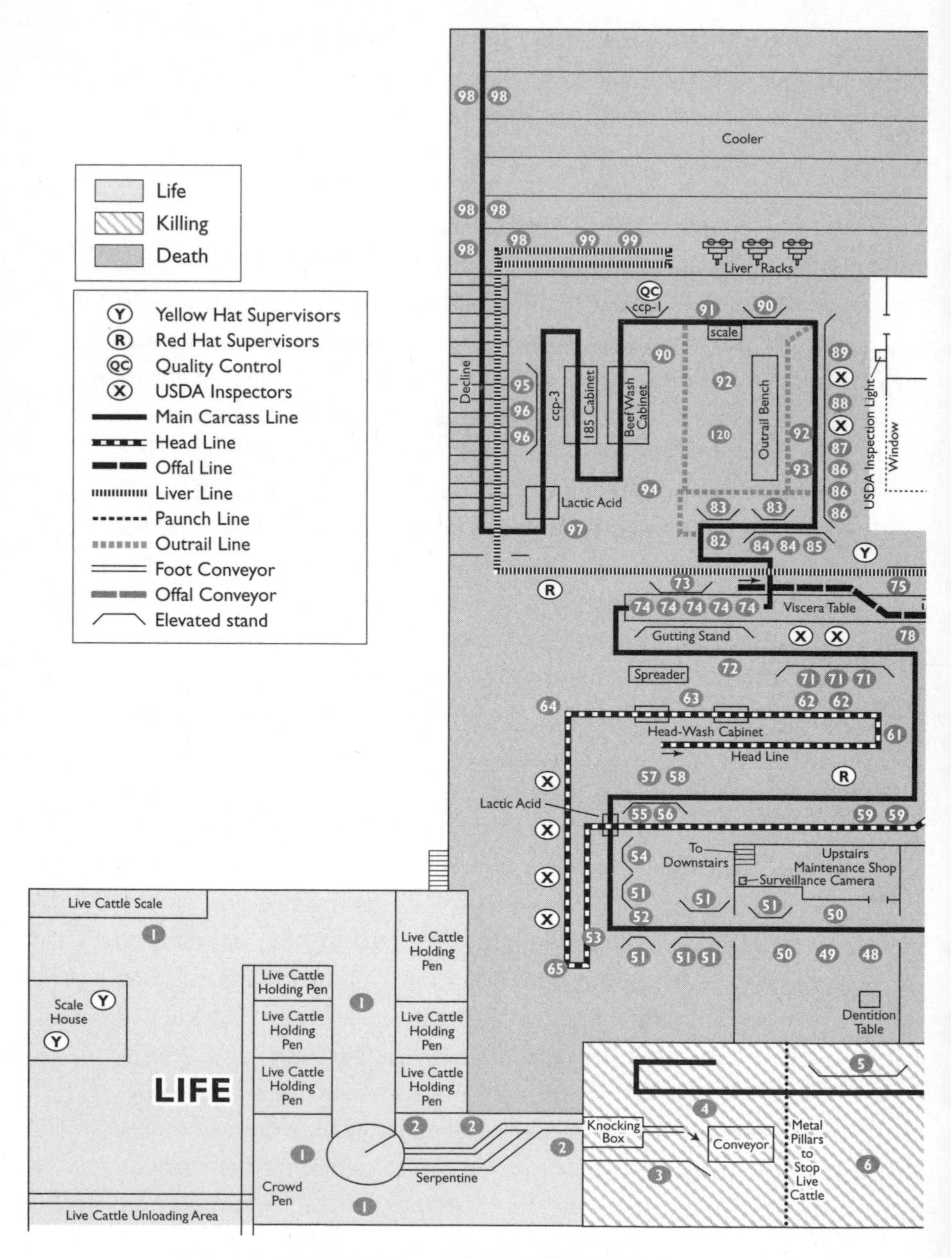

Fig.8 Divisions of Life, Killing, Death

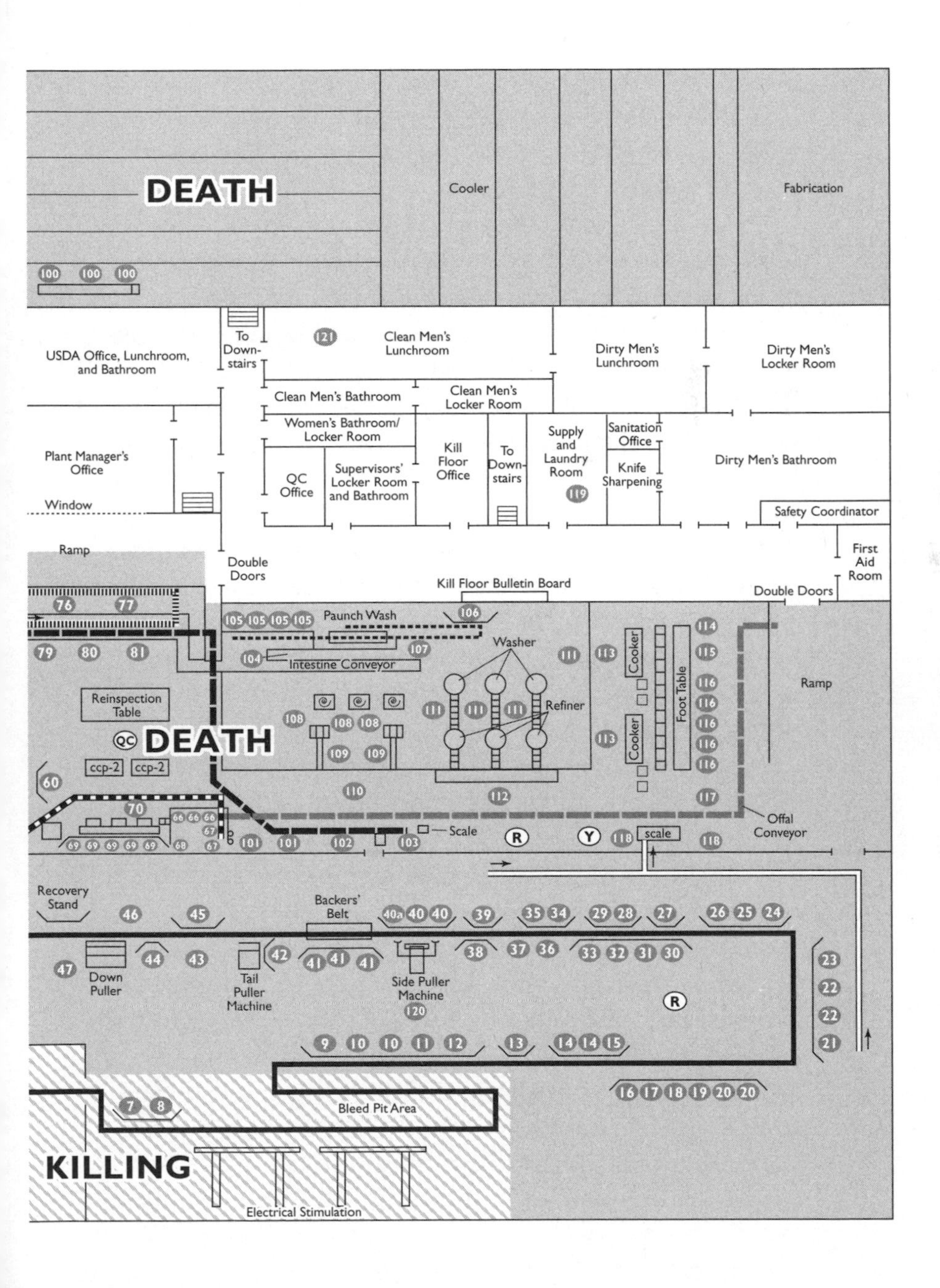

DEATH
Cooler
Fabrication
100 100 100
USDA Office, Lunchroom, and Bathroom
To Down-stairs
121
Clean Men's Lunchroom
Dirty Men's Lunchroom
Dirty Men's Locker Room
Clean Men's Bathroom
Clean Men's Locker Room
Women's Bathroom/ Locker Room
Plant Manager's Office
QC Office
Supervisors' Locker Room and Bathroom
Kill Floor Office
To Down-stairs
Supply and Laundry Room
119
Sanitation Office
Knife Sharpening
Dirty Men's Bathroom
Window
Safety Coordinator
Ramp
Double Doors
First Aid Room
Kill Floor Bulletin Board
Double Doors
76 77
105 105 105 105
Paunch Wash
106
79 80 81
104
Intestine Conveyor
107
Washer
111
113
Cooker
Foot Table
114 115 116 116 116 116 116
Ramp
Reinspection Table
108 108 108
111 111 111
Refiner
QC DEATH
109 109
113
Cooker
ccp-2 ccp-2
60
110
112
117
70
Offal Conveyor
66 66 66 67 67
69 69 69 69 69 68
101 101 102 103
Scale
R
Y
118
scale
118
Recovery Stand
46
45
Backers' Belt
40a 40 40
39
35 34
29 28
27
26 25 24
47
Down Puller
44
43
Tail Puller Machine
42
41 41 41
Side Puller Machine
120
38
37 36
33 32 31 30
23 22 22 21
R
9 10 10 11 12
13
14 14 15
16 17 18 19 20 20
7 8
Bleed Pit Area
KILLING
Electrical Stimulation

as well), and shoots the cow. If the shot is good, the cow falls to the floor, and the maintenance crew hoists the fallen cow with a Bobcat lifter onto the green-plastic conveyor, where it is re-shackled and raised onto the line.

A variation of this scenario is when the knocker has not successfully stunned the cow and the failure goes unnoticed by the indexer. The cow arrives, hoisted by its hind leg, at the presticker and sticker stand still conscious, kicking and swinging wildly. At this point the presticker and sticker face a decision. Behind them, attached to a pole, is a red button that will stop the entire production line. If there is a USDA inspector within close proximity, the presticker or sticker will push the button and wait until the indexer can come to properly stun the cow with the handgun. If there is no inspector nearby, the presticker and sticker will attempt to cut the animal, but the cut is often imperfect since the risk of getting hurt when cutting into a conscious cow is very high. It is possible, given the imperfect cut, that the cow will then continue through the electrical stimulation and bleed pit area, still conscious. The tail ripper (circle 9), the first leggers (circle 10), and the bung capper (circle 11) will begin cutting into the cow's tail, right rear leg, and anus, respectively, while the cow is still sentient.

Because these workers stand on a platform elevated ten feet above the kill floor, the head of the cow is invisible to them, and they are unaware that they are cutting into a sentient animal. The reaction of the cow to the pain of being cut into is visible only from the floor and observable from the movements of the cow's head and eyes.[2]

These transgressions make clear the precision of the boundaries separating life from death on the kill floor. There are certain points before which cattle must be alive and certain points beyond which they must be dead. The area in between those two points extends for the approximately fifty feet separating the knocker and the stickers, and it is the designated space for dying.

As figure 9 illustrates, it is possible to tabulate how many workers out of a workforce of approximately eight hundred see the cattle alive, are actively involved in killing the cattle, or have a sight line to live cattle or the killing of the cattle. In the separation between life and death, the majority of slaughterhouse workers operate in the zone of death. Only a few see the cattle while they are alive or are in the process of being killed, and an even smaller number are actively involved in the killing. Furthermore, the act of killing itself is divided into more stages, which are also out of sight of one another. As I describe in greater detail below, space and labor on the kill floor, itself a department isolated from the rest of the slaughterhouse, are organized to segregate and quarantine the killing.

Clean/Dirty

The separation of space that has the greatest impact on the organization of work on the kill floor is that between the "clean side" and the "dirty side." The physical point dividing these two zones occurs at the down puller (circle 47), where a large hydraulic machine rips the hide off the suspended cattle. The area where everything is done previous to this point is referred to as the dirty side, and the area beyond this point is referred to as the clean side (fig. 10). Sometimes the clean side and dirty side are also referred to as the hide-off and hide-on areas.

The primary logic behind this distinction is food safety. The cattle come into the plant with feces, vomit, and ingesta on their hides, and from a food-safety perspective the challenge is to take the hides off them while minimizing the transfer of

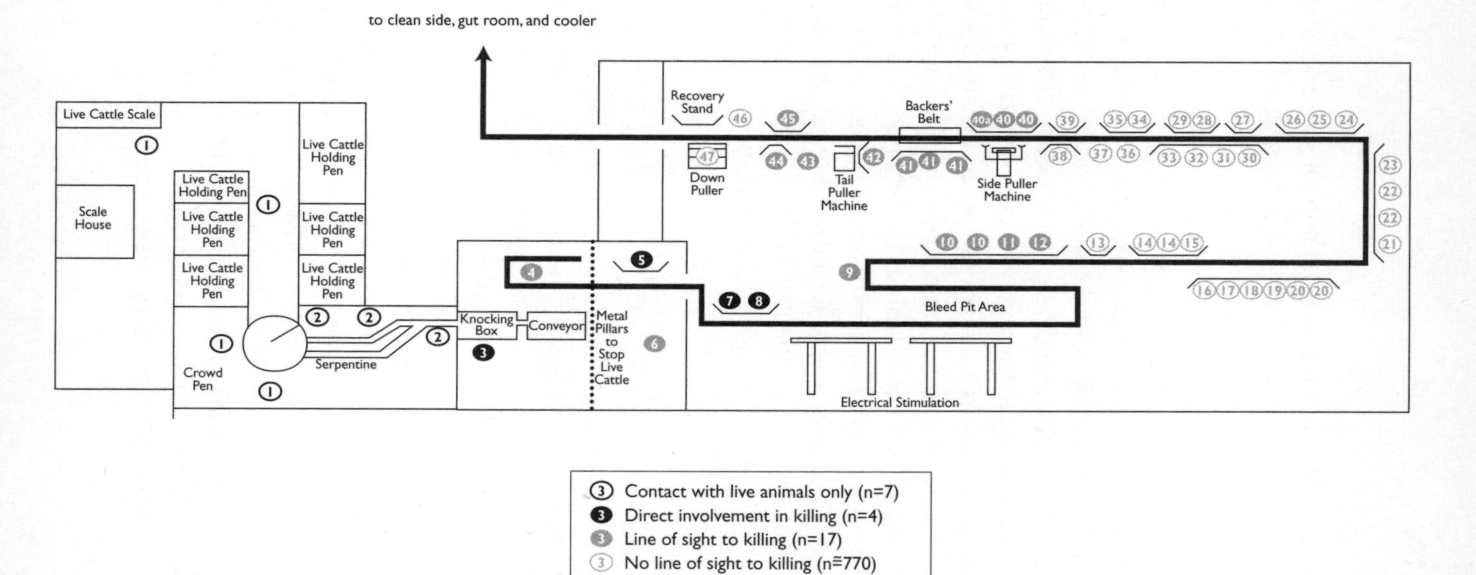

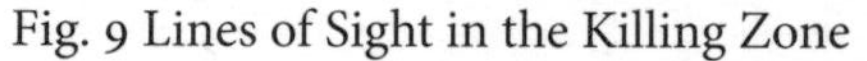

Fig. 9 Lines of Sight in the Killing Zone

these contaminants to the flesh underneath. Three large hydraulic machines rip the hides from the flesh. In order, these are the side puller (circle 40), the tail puller (circle 42), and the down puller. From the tail ripper (circle 9) on, the workers before and between these machines are tasked with making the preparatory cuts and incisions that prep the hide for the heavy ripping performed by these machines. Often this involves using a hand knife or an air knife to cut between the hide and the skin, a job that sometimes requires the worker to pull back on the hide or otherwise make contact with it. From the standpoint of food safety, the hands, clothing, and tools of these workers then become possible conduits for contamination of the meat.

One partial solution to the problem of cross-contamination is to segregate the workers and their equipment from the rest of the plant. These quarantined workers are known as the dirty men, and unlike their counterparts on the clean side, who wear white helmets, they wear gray helmets. The dirty-side workers have their own toilets, showers, locker room, and lunchroom. On the safety map posted in the kill floor hallway showing emergency exit routes, these rooms are labeled "Dirty Men's Bathroom," "Dirty Men's Shower," and "Dirty Men's Lunchroom." At the end of the day, the dirty-side workers place their clothing in a separate hamper from the one the other workers use. On the entire dirty side, there are only two female workers: the bung stuffer (circle 31) and the Whizard knife belly trimmer (circle 40a). There are no separate facilities for the "dirty women," and they share a bathroom with the "clean women." The male workers on the clean side have their own facilities, labeled "Clean Men's Bathroom," "Clean Men's Shower," and "Clean Men's Lunchroom."

The physical separation between the clean and dirty

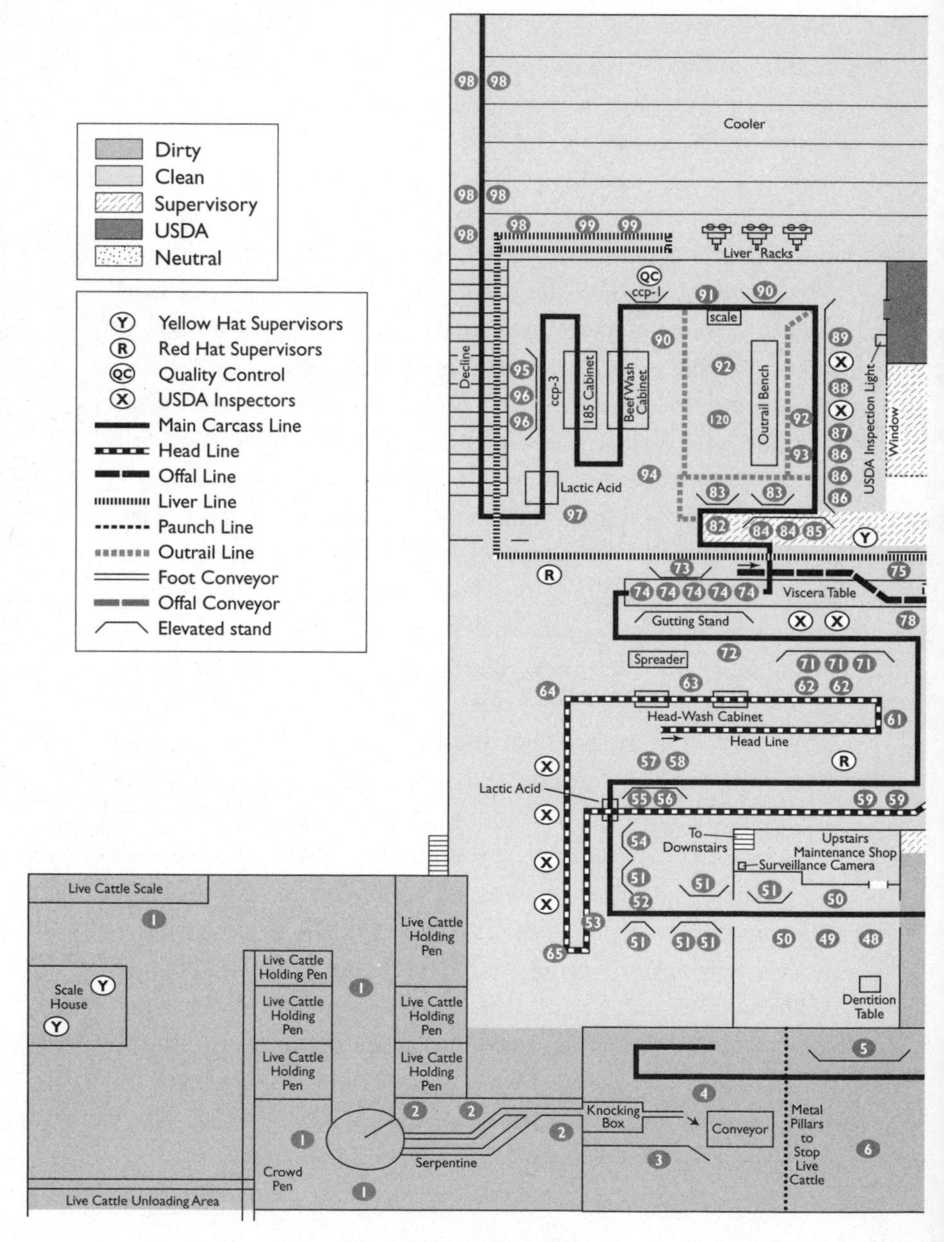

Fig. 10 Dirty and Clean

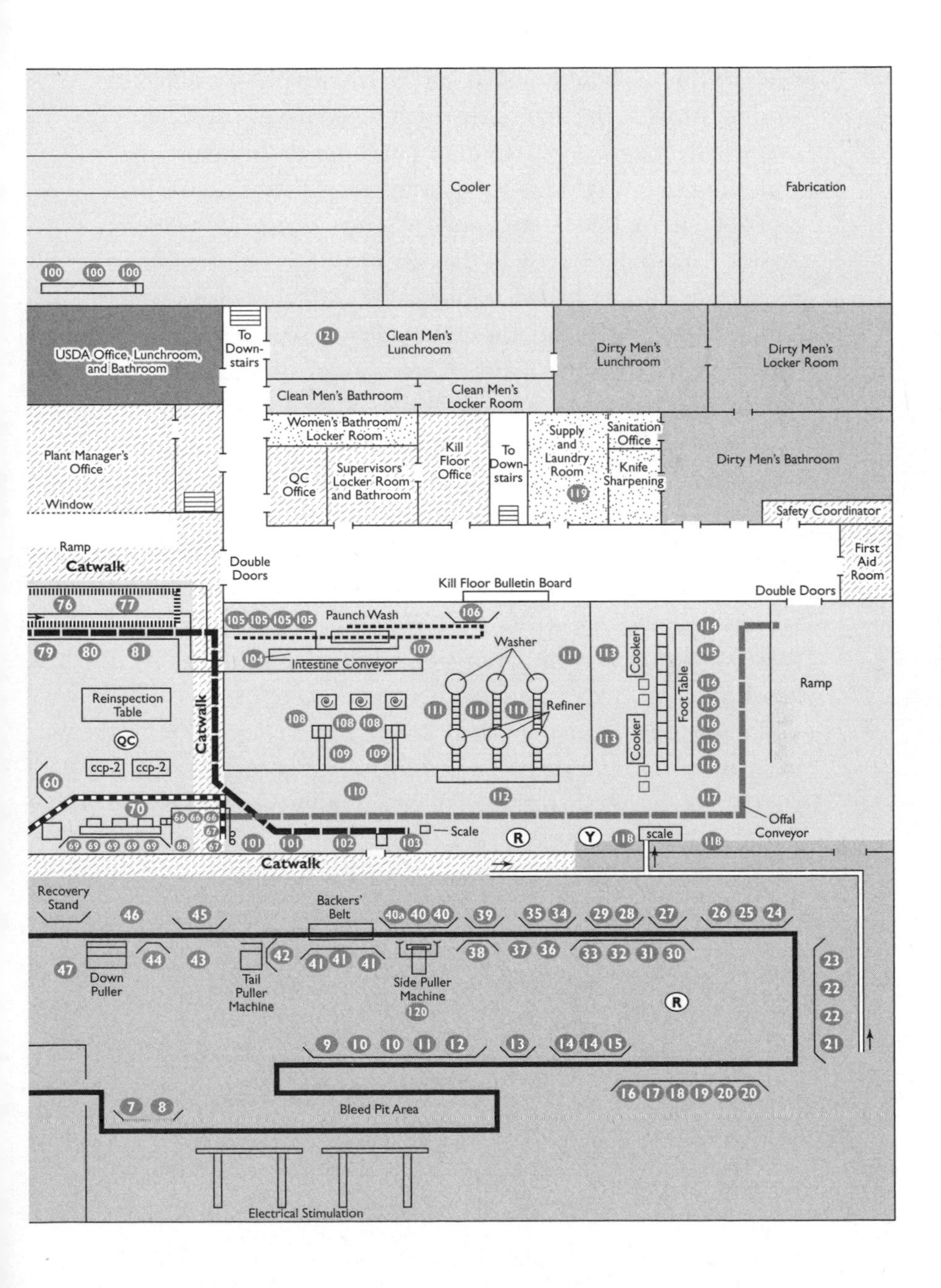
Cooler
Fabrication
USDA Office, Lunchroom, and Bathroom
To Downstairs
Clean Men's Lunchroom
Dirty Men's Lunchroom
Dirty Men's Locker Room
Clean Men's Bathroom
Clean Men's Locker Room
Plant Manager's Office
Women's Bathroom/ Locker Room
Kill Floor Office
To Downstairs
Supply and Laundry Room
Sanitation Office
Knife Sharpening
Dirty Men's Bathroom
QC Office
Supervisors' Locker Room and Bathroom
Window
Safety Coordinator
Ramp
Catwalk
Double Doors
First Aid Room
Kill Floor Bulletin Board
Double Doors
Paunch Wash
Washer
Intestine Conveyor
Cooker
Foot Table
Ramp
Reinspection Table
Refiner
Catwalk
QC
ccp-2
ccp-2
Scale
scale
Offal Conveyor
Catwalk
Recovery Stand
Backers' Belt
Down Puller
Tail Puller Machine
Side Puller Machine
Bleed Pit Area
Electrical Stimulation

sides of the slaughterhouse is further reinforced by different work schedules. The first cattle reach the knocking box at 6:30 every morning, and it is another half hour or so before they pass from the dirty side to the clean side. Dirty-side workers must be on the line at 6:30 A.M., whereas clean-side workers do not start until 7:00, or, in the case of workers farther along the line, even later. Likewise, dirty-side workers finish work a half hour or more before the clean-side workers.

As with the other pairs of separations on the kill floor, the division between dirty and clean, so ideal from the planner's point of view, is constantly transgressed. Certain categories of workers—supervisors, sanitation workers, quality-control workers, USDA inspectors—routinely and repeatedly cross between the dirty and clean sides without changing their clothing or cleaning their boots. Workers from the dirty side often use the microwaves in the clean-side lunchroom to heat up their lunches since the clean-side workers take their lunch break a half hour later than they do, and the microwaves in the dirty-side lunchroom are often used to full capacity. Workers from the clean and dirty sides mingle freely outside in the parking lot during their shared fifteen-minute morning break, eating together at the taco trucks parked there.

Indeed, what was intended as a functional separation from the standpoint of the kill floor and food-safety planners may have more value as an experiential marker. If the presence of an animal's hide is a proxy for clean and dirty from the standpoint of the food-safety planner, it also operates as a proxy for animal and carcass from the standpoint of what the workers see. To watch the movement of a cow from the chute to the down puller is to witness the transformation of a creature from fully animal to carcass. In the chutes, each of the cattle has its own unique characteristics: breed, sex, height, width, hide pattern, level of curiosity, eyes, horns, sound of bellow. From a phenomenological standpoint, after the cattle are stunned, shackled, and suspended upside down in the air, the entire process seems geared to stripping them of these unique identifiers in order to begin the process of turning living animals into homogeneous raw material. This process will not be completed until the animal leaves the fabrication department in boxes, but it is here, at the earliest stages, where the change is most dramatic.

The tail ripper begins by cutting off the bottom third of each cow's tail, then making an incision from the anus to the inside of the right back leg. The first leggers take over from there, deftly cutting a pattern in the hide of the right back leg, pulling back the hide to expose the first glimpse of the pearly white flesh underneath. And so it goes along the line, each worker making his or her small cut or pattern in the hide, gradually pulling it back to reveal more and more of the meat that drives the entire process. Hoofs are cut off with huge, handheld mechanical shears, which carry the warning "Danger: Risk of Amputation."

Air knives whiz around the rump area, pulling back the hide just enough for the breed stamper (circle 30) to leave in blue ink that one mark of identity that persists all the way to fabrication: "A" for Angus, cattle with a pure black coat; "H" for Hereford, those with a reddish coat; "C" for mixed Angus and Hereford, which have a black coat with substantial spots of white mixed in. Some breeds, such as the leggy black-and-white-spotted dairy Holsteins so iconic as a typical representation of "the cow," do not even merit, from a culinary perspective, the distinction of a breed stamp; they pass through unmarked.

Two workers (circles 36 and 37) stand in a pit of blood,

sticking a gloved hand into the cow's nostril, stretching it out, then severing it with a single cut from a hand knife. The ear is next, its tip grasped in one hand before it is sliced neatly in one cut. A narrow but forceful geyser of blood often spurts out, sometimes hitting the worker in the eye. The ear is tossed into a tub, which by the end of the day will hold more than five thousand ears.

Finally, the animal is prepared for the side puller. Through a series of cuts along the assembly line, its belly area has been completely stripped of hide, which now hangs like an open robe, exposing an obese roll of white flesh. Two men take one end each of this robe and pull it into a foot-long metal clamp before depressing a button that causes the clamp to close around the hide. One of the men turns to a control panel and activates a lever that causes the hydraulic arms to pull back, ripping the hide from the underbelly of the cow and leaving the hide attached only to the animal's backside.

Three backers (circle 41) step onto a conveyor belt that moves at the same speed as the passing cattle and make a "pocket" in the center of the cow's back with air knives. This pocket offers an insertion point for the arm of the second hydraulic machine, the tail puller. The backers work in almost perfect synchrony, starting on the right side of the moving cattle and using the left hand to squeeze the air knife trigger and carve a space between the hide and flesh, before stepping around—still in unison, for there would be no room otherwise—and switching the air knife to the right hand to carve a space on the left side of the cattle, thus creating a passageway between the hide and the flesh on the cow's back. Once the passage is created between hide and flesh the backers immediately step off the belt and go over to three sinks, where they splash some water on their aprons and then dip their knives into a container of water heated to 185 degrees. Then they step back again for another round of their elaborately choreographed dance, darting in and out among the moving carcasses.

At the tail puller, the worker inserts a thick metal arm called a banana bar into the space just carved by the backers and pushes a button, and the metal arm swings forcefully upward, stripping the hide from the midpoint in the back all the way up to the tail. Now the hide drapes like a cape over the back and head of the cow, hiding them in its folds. The top half of the cow is completely denuded now, white and pale under the glow of the halogen lights.

Then, at last, the cow reaches the down puller, where a worker uses a hook to pull the draping hide into a long circular piece of metal that at once functions as clamp and roller. Once the length of the hide is caught in the clamp, the worker throws a lever and the roller spins furiously, ripping the rest of the hide down off the head of the animal. When the hide is completely off, the worker pulls a second lever, and the roller spins back the other way, depositing the hide down a chute directly below. Now the cow's appearance is surreal: a pearly white creature with bulging eyeballs, broken teeth, and perforated head, which drags along the metal roller that has just stripped it of its hide.

Sometimes at the down puller the force of the clamp roller will be so extreme that the tendons of the hocks will break, and the cow will fall with a thud to the blood-soaked floor. If this happens, it has to be hoisted by a winch and re-railed onto the line. Sometimes the down puller will not exert enough force or will catch on a horn that has not been taken completely off, and this will leave the entire hide hanging from the head of the cow dragging on the floor. When this happens, the down-puller operator sounds a special horn, without stopping the production line, and a supervisor comes with a knife to remove the hide by hand. This is tedious work, and time is limited, for the cow has now trespassed into the clean side, and the hide must be completely removed before the carcass turns the corner past the vacuum cleaners (circle 51) to enter the prewash cabinet (circle 54).

Absent hoofs, ears, lips, horns, and hide, the cow now begins its journey through the clean side of the kill floor. Here the de-animalization continues with the severing of the cow's head (circle 58) and tail (circle 73), removal of the intestines and internal organs (circle 74), and vertical splitting of the carcass through the spinal cord (circle 83) to create the half-sides that will hang in the cooler before going on to fabrication. Although these actions further decrease the visual resemblance of carcass to animal, they also produce a massive concentration and accumulation of smaller body parts, an incredible visual spectacle.

Once severed, heads are hung on moving hooks that constitute a separate work line known as the head chain. No longer attached to a body, these heads float at chest level from the head severer to the head flusher (circle 61), then to workers who remove the tongues and hang them on hooks next to the heads (circle 62). The line continues through two cabinets from which water is sprayed out of nozzles, like a miniature car wash (circle 63), before it turns past a line of four USDA inspectors, who make small incisions and examine the glands in the heads for signs of disease or abnormality. Once past the inspectors, the head line doubles back on itself and angles steeply upward, crossing over the main line of carcasses, which are just exiting the prewash cabinet, still in possession of their heads. Suspended on hooks, the heads and tongues travel high above the kill floor, almost level with the catwalk positioned some thirty feet above the floor. A metal drip pan hangs about three feet below them, catching most, but not all, of the blood that continues to flow. Then the head line angles downward again, entering the head-table area, a combination of a raised stand and a conveyor belt, where eleven workers (circles 66 through 69) cut off lips and extract as much flesh as possible from the heads. By the time they are finished, the jaw has been detached from the skull, and each transformed into a polished ivory object that exits the kill floor on a rubber conveyor belt and falls into an open-bed truck for transport to a rendering factory. The flesh extracted from the head and jaw is divided into three categories: head meat, cheek meat, and lip meat, all of which will be boxed in sixty-pound units (circle 70), placed on a conveyor to the warehouse, and sold.

As with the heads, so too with the livers: they are gutted (circle 74), inspected by USDA officials working at the viscera table (circle 78), then placed on hooks on a separate chain (circle 77) that angles upward, high above the kill floor, before descending through a hole in the wall down the decline to the cooler, where they are hung on carts to cool (circle 99) and then be packed (circle 100). The tails, hearts, and weasand (the lining of the esophagus, used in hamburger meat) share the same offal chain once they have been removed from the animal. This chain crosses high over the viscera table (circle 78), runs parallel to the wall separating the gut room from the kill floor, then angles downward to the offal packer (circle 102) and the tail washer and packer (circle 103), who remove them from the hooks, pack them, and weigh them.

The clean side of the kill floor, then, is vastly more differentiated than the dirty side, where there is only one main line. In addition to the main chain of suspended white carcasses, three additional chains crisscross the clean side of the kill floor at different heights and angles, sometimes rising high above the field of vision and at other times angling down to chest height. Each of these chains carries specific body parts, concentrated and aggregated. On one line, a row of pale moving heads stripped of their skin, eyeballs bulging out of their sockets, and tongues hanging out of their mouths. On another, an alternating row of tails with muscles still reflexively twitching, sinewy hearts, and dangling weasands. And on yet another line, a row of weighty, elephant ear–shaped livers, moving steadily in solemn procession.

The visual confrontation with the head line is shocking: taking in head after head gliding above the kill floor in what seems an endless succession, one is able to grasp, with a vividness that exceeds even what one sees on the dirty side, the sheer, staggering *volume* of the killing. The disembodied massing of that one small portion of the animal, which of all the body parts continues to refer most unambiguously to life—the face—offers a haunting image of vast destruction. Yet in the efficient homogenizing of the slaughterhouse, even this most individualized of body parts will soon be reduced to the common, salable categories of head, cheek, and lip meat, the gleaming, stripped skulls and broken jaws sold to a rendering company.

It is different with the lines of tails, weasands, hearts, and livers. Unlike the line of faces on the head line, these body parts are anonymous and interchangeable. Although the muscles in the tails may still twitch reflexively and the livers give off a visible heat, it takes a concentrated act of the imagination to reconstruct the whole animal from these bits and pieces.

Main Line/Auxiliary

The body parts separated from the carcasses require auxiliary production lines all their own. Central to these are the head table, where meat is taken from the skull and jaw; the offal-packing area, where hearts, weasands, and tails are weighed and put into boxes; the gut room, where stomachs and intestines are flushed and washed; the foot room, where hoofs are washed, cooked, and packed; and the pet-food room, which will be considered below under "Upstairs/Downstairs." Appendix B, "Cattle Body Parts and Their Uses," offers an overview of the astonishing array of fields and products for which these noncarcass body parts are used.

The distinction between main-line and auxiliary work also serves as a proxy for a gender distinction in kill floor work. Of the twelve women who work on the kill floor, only two work on the main line; the rest work in one of the auxiliary operations. Main-line operations become coded as "real" work, "men's" work, and auxiliary operations are coded as "secondary" work, suitable for women and male workers who are injured or otherwise unfit for the vigor of main-line work. The actual physical demands of the job are irrelevant; what matters is the image of central versus peripheral work. A slaughterhouse would still be a slaughterhouse even if it did not process feet and offal; its main product is beef. The irony of the distinction is that these "auxiliary" operations are highly profitable for the slaughterhouse. Without them, an industrialized slaughterhouse would no longer be economically competitive.

Of the women who work in the auxiliary operations, five work together, side by side in the foot room, trimming the fat and discolorations off the cooked feet before they are packed into boxes (circle 116). Whether by intention or accident, the trimming stand is set up so that the women work with their backs facing the main ramp that leads from the hallway through the foot room to the dirty side. Near the end of the day male supervisors often gather at the concrete railing that separates the ramp from the production floor. Sitting on the ramp with their legs dangling over the side, they openly exchange comments about the relative degree of attractiveness of the foot trimmers' bodies, jeering, laughing, and leering when the women have to bend over to pick up feet out of a metal tub to pack them into the boxes.

If auxiliary work creates predatory spaces where workers can become the object of an unwelcome gaze, it also creates areas where some of the dirtiest work of the kill floor takes place out of sight, creating yet another layer of invisibility within the kill floor itself. The gut room is one such place (circles 105–111). After the cattle paunches (stomachs and large intestines) have been eviscerated by the gutters, they pass via a metal conveyor belt from the viscera table through a small gap in the wall into the gut room. Here stomachs are separated from the large and small intestines. The stomachs are pushed down another conveyor belt to the paunch-opening room, where four workers (circle 105) labor in a tiny cell no larger than six feet by twelve, cutting open the stomachs, spilling out their half-digested contents onto a metal table, then hanging the stomachs on hooks that carry them through an automated water wash and on to a trimmer (circle 106), who cuts the stomachs off the hooks and puts them into a large circular vat, where they are washed in powerful chemicals (circle 111).

The miniscule room occupied by the paunch openers has no ventilation, though a rusty oscillating fan sits at the opening. When on, it succeeds only in recirculating the fetid air within the room. Concrete walls enclose the workers on all sides;

the only opening is the small, three by two–foot space through which the conveyor, bearing an endless procession of stomachs, passes. The walls seem to squeeze in on the space, which is further compressed by the way they rise twenty-five or thirty feet before ending in translucent roofing material, which lets some natural light in without allowing the workers to see out.

The paunch-opening room makes the rest of the kill floor smell benign by comparison. When the stomachs are cut open, a thick odor escapes; it is like a combination of the acrid smell of vomit and the sulfuric stench of rotting eggs. Partially digested dark-green feed dotted with intact kernels of bright yellow corn spills out onto the metal cutting table, and the workers use their gloved hands to scoop out the rest, pushing it across the table and down a hole to the hopper, a large funnel-like machine on the floor below. Often a small circular piece of metal about the size of a pack of gum will be among the contents of the stomach: a cow magnet, fed to the cows to attract any foreign metals they might accidentally ingest.

The air in the paunch-opening room is warm and humid with the heat and gases of thousands of four-chambered bovine stomachs. An hour or two into their work, the paunch workers are drenched in sweat, adding a human stench to the already heady olfactory mix. The workers absorb this smell into their very skin: when they walk through the locker rooms, even after showering at the end of the day, the other workers hold their breaths, some less discreetly than others. The stink of the paunch worker is distinct even within the broader stench of the kill floor, and the poisonous smell of the paunch-opening room creates its own relatively surveillance-free zone. As long as the workers keep hanging the stomachs on the paunch hooks, they can count on receiving only rare visits from plant supervisors and USDA inspectors.

Although more spacious, the rest of the gut room is no less dirty. The small intestines are taken off the conveyor belt by three workers (circle 108), who bend over at the waist to thread them onto metal coils, which have openings for streams of water to flush out the small intestines. Because it is pressurized, this water punches small holes in the intestines, causing bits of white intestine to fly through the air, hitting workers in the face and body. By the end of the day, the workers are covered with pieces of intestines and their contents. Even their hair is flecked with innards; the angle of the work allows the small bits to be flung into their hard hats. Except for the supervisor specifically responsible for the gut room, plant managers and USDA inspectors rarely walk through this room. A space unto itself within a space unto itself, the gut room segregates the work that is dirty even by the standards of the kill floor.

Upstairs/Downstairs

The divisions between outside and inside, alive and dead, clean and dirty, and main line and auxiliary all pertain to the upper level of the kill floor, where the central operations take place. But the distinction between upper and lower levels represents another important division, both physical and social. Most of the lower level of the kill floor is divided into rooms that mechanically or chemically support the operations on the upper level.

Although maintenance has a small shop area on the upper level of the kill floor, its main work and parts-supply area takes up most of the lower level. Maintenance workers wear purple hard hats and, unlike the line workers on the kill floor, dark-blue uniforms provided by Aramark. There are three shifts of maintenance workers at the plant. The day shift handles breakdowns in equipment that occur during operations. The evening shift checks equipment at day's end and readies it for the next day. The night shift works on long-term projects, such as welding chutes, changing sprockets, and creating new or modified workstations.

All maintenance workers carry radios that enable them to communicate with one another across the plant floor as well as with the kill floor supervisors and managers. During the day, the code "Mayday" is used to signify a breakdown somewhere on the line that causes the entire line to stop. If there is a problem with the chain that descends from the kill floor to the cooler, for example, someone will call on the radio, "Mayday at the decline. Mayday at the decline." Within minutes about seven or eight maintenance workers will converge on the steps leading down to the cooler, assess the situation, and set to work correcting the problem.

Maintenance workers have their own break room on the lower level of the kill floor, and, unlike the dirty- and clean-side break rooms used by the line workers on the upper level, which have microwaves, this room has only a hot plate. Taped to the outside of a dirty white refrigerator is a piece of plain white paper with large boldface print commanding, "Leave other peoples lunch and pop alone. Quit stealing other peoples lunch." It is not uncommon for those walking past the maintenance lunchroom to see workers reading the newspaper, talking, leaning against the wall, or eating. Unlike line workers, whose day consists of monotonous, unending demands, maintenance workers operate in constant standby mode, with long stretches of free time punctuated by unpredictable crises.

In addition to the maintenance shop, parts room, and break room, several other rooms on the lower level support the operations of the kill floor. These include the air-compression room, which provides power for all the air tools on the upper level, including the air knives and the knocker's gun; the boiler room, which heats the water to 185 degrees and pumps it upstairs for sterilizing knives and other equipment; the room with the electrical panel; the chemical-storage room, where chemicals of all sorts are stored in barrels; the hook room, where a single worker takes the hooks that hold the carcasses on the line and washes them before they are sent back into use; the hopper room, where the ingesta of the cattle are dumped after their stomachs are slit open by the paunch openers; and the box room, where two workers construct cardboard boxes and place them on a conveyor that moves them to the offal-packing area on the upper level (circle 112).

These rooms exemplify the segmentation created by the division of labor. Workers in the box room, for example, enter the slaughterhouse on the lower level and go into the upper level only to use the clean men's locker room and lunchroom. They never enter the main operation space of the kill floor and in the course of a typical workday have no contact with anything resembling an animal. All day long, they fold boxes and place them on a conveyor belt. Likewise, the worker in the hook-cleaning room sees only pieces of fat and specks of blood on the hooks that circle down the rail and clang against one another with tremendous noise.

A notable exception to the abstract environment of the lower level of the kill floor is the pet-food room, which sits off a corridor across from the chemical-storage room and next to the hopper room. Two workers labor in this room. One, an employee of the slaughterhouse, wears a gray helmet and stands on raised plastic grating in front of a long metal table. The table is pushed against a dirty white metal wall and sits just under an angled pipe about a foot and a half in diameter that protrudes through the wall. This pipe is connected to a metal funnel located next to the offal hanger (circle 77) on the upper level. All day long the offal hanger throws four different cattle parts down this funnel, and all day long they come shooting out of the pipe at the other end, landing with a dull thump: lungs, windpipes, kidneys, and rejected livers. The gray-helmeted worker uses a long metal rod with a hook at the end to latch onto the organs and fling them into one of many square gray tubs in a semicircle around him. The tubs are about eight feet square and five feet high, and on their sides in black military-style block letters are the words "NOT FOR HUMAN CONSUMPTION." There is one tub for livers, two for lungs, three for windpipes, and four for kidneys. All day long, this worker stands with the four foot–wide metal table separating him from a blank white wall flinging organs into tubs. Those that land on the floor are picked up throughout the day and put into their intended receptacles.

The second worker is employed by a medical lab that pays the slaughterhouse to give him access to this room. He stands on a raised webbed platform at another long metal table; this one has a vertical metal screen running down its length studded with sharp metal hooks, about fifteen in all. Sometimes out of the pipe in the wall an oblong gray mass shoots that is not a lung, kidney, windpipe, or liver. When that happens the white-helmeted worker walks over, picks up the object, and carries it back to his worktable, where he takes out a knife and cuts into the gray mass. There will be a fetus inside, with smooth, slick skin, and clearly marked hide patterns. Raising the fetus up by the neck and hind legs, the man swivels to the vertical metal screen and pushes the fetus's mouth onto one of the protruding hooks. Releasing the neck so that the body now hangs by the mouth, he uses two hands to stick another hook into the fetus's anus. The fetus now hangs suspended by its mouth and anus, and the worker makes an incision in the neck area, bringing a bottle with a straw cut at an incline up to the incision.

He then shakes and massages the body of the fetus, coaxing blood into the waiting bottle. The shaking becomes more vigorous as less and less blood remains in the body. Finally, when there is no more blood to be had, the man pulls the bottle from the incision, caps the straw, and nestles the bottle inside the chipped ice of a blue ice chest for later use in medical production. Once bled, the fetus is deposited in a gray circular barrel on top of other bled fetuses.

The downstairs space, then, is marked by the juxtaposition of rooms that provide purely mechanical and chemical support with rooms that extend the protracted confrontation with the biological taking place upstairs. Side by side, in adjacent rooms, workers can either spend their entire time at the slaughterhouse without more than a passing glance at a cow or any part of a cow or they can spend the whole day coaxing blood out of cattle fetuses.

Supervisory/Production

In addition to the division of space through physical mechanisms such as the placement of walls and doors, the use of overhead rails and chains, and the separation between upstairs and downstairs, the kill floor is also divided into hierarchical spaces that have as their physical manifestation the color of hard hat worn by a worker. I have already noted that clean- and dirty-side workers are differentiated by white and gray hard hats. This distinction allows everyone to see at a glance whether a worker is "out of place" or "out of line," common figures of speech that take on literal meanings in the context of the kill floor. In addition to the gray/white hard hat distinction, some production workers wear blue hard hats and some orange hard hats. These are, respectively, the dentition and sanitation workers.

The dentition workers (circles 48 and 49) are stationed just inside the clean side, right after the down puller. Their job was created after the discovery of bovine spongiform encephalopathy (BSE), or "mad-cow" disease, in the United States. Because the USDA determined that cows aged thirty months or more are at a higher risk for BSE, slaughterhouses across the United States are required to label and market cattle of this age separately from other cattle. On the kill floor, these higher-risk cattle become known as thirty-month cattle.

The task of the blue-helmeted dentition worker is to swing the lower jaws of cattle open as they pass and determine, by examining the teeth, whether the cattle are older or younger than thirty months. Cattle with two or more permanent teeth are older than thirty months; cattle with fewer than two permanent teeth are younger. When a "thirty-month" cow (that is, aged thirty months or older) passes by, the dentition workers stamp the number "30" in blue ink on its shoulders, put a cork in the hole in its head to prevent more brain material from falling out, and clip a rectangular red paper tag labeled "30 months" through its eyelid and shoulder.

The physical demands of this job are less onerous than those of many others on the line, and blue-helmeted workers are given a radio that allows them to monitor plant communication and call for backup if there are too many consecutive thirty-month cattle to deal with. Nonetheless, the job is viewed with consternation. The clear-cut guidelines about identifying a cow's age do not always correspond with the reality of cattle that have had teeth knocked out or with the vomit, ingesta, and blood that often obscure a clear view of the inside of the mouth. In addition, the monotony of looking into the jaws of twenty-five hundred cattle at the rate of one every twelve seconds for nine to twelve hours a day can lull dentition workers into missing a cow with two or more permanent teeth. If this missed cow is later caught by USDA inspectors, the dentition worker is summarily fired or, if he is lucky, placed on a three-day suspension and moved to a different job.

Sanitation workers, who wear orange hats, have some of the most physically grueling work on the kill floor. Their job is to move around the plant emptying barrels of fat into rendering chutes, shoveling fat that has fallen onto the floor, and pushing accumulated blood across the floors to the drains. They must take care not to touch the moving cattle and cattle parts with any part of their bodies or equipment, and they risk being fired or suspended if they do.

In addition to the division between white, gray, blue, and orange hard hats among production workers, there is also a distinction between different types of nonproduction workers. Floor supervisors wear red hats; their assistants, known as utility men, wear yellow hats; and quality-control workers wear green hats. In contrast to production workers, who wear their own clothing to work, these workers are supplied with uniforms and a uniform-cleaning service.

Red-, yellow-, and green-hat workers have a freedom of movement not available to those with gray and white hats. Gray- and white-hat workers are tied to a specific spot on the line and cannot move from that spot unless someone relieves them. Workers with red, yellow, and green hats have the freedom to roam the kill floor and, significantly, can use the toilets whenever they need to. The mobility enjoyed by red-, yellow-, and green-hat workers extends to nonproduction spaces on the kill floor as well. Supervisory staff have access to all the break areas available to the line workers (lunchrooms, locker rooms, and bathrooms) as well as to office and break spaces that are off-limits to line workers. Red-hat workers have their own locker rooms and bathrooms, as well as their own office space. Green-hat workers have their own office space, where they fill out and file paperwork.

Additionally, workers with red, yellow, and green hats are all equipped with, and expected to use, radios that keep them in communication with one another and with plant managers. These radios create a division in communication that allows a small group of workers access to key information not available to line workers (such as how many cattle are going to be slaughtered that day, what the likely quitting time will be, and where various USDA inspectors are located at any given time). Officially, radios are used to communicate about problems that arise in the plant, line workers who are not performing their jobs satisfactorily, and the location and activities of USDA inspectors (who do not have access to radios). Unofficially, these radios are also used for informal communication, thus creating and reinforcing a sense of difference and superiority of nonline over line workers.

Radios, however, also function as a technology of control. The two kill floor managers use the radios to monitor the whereabouts and activities of red-, yellow-, and green-hat workers. When one of these is called by a manager on the radio, he or she has about five or six seconds to respond before the ire and suspicion of the manager is raised. The kill floor managers have the freedom of the entire plant, as well as occupancy of the manager's office, a large, nicely furnished room in a corner overlooking both the trim rail (circles 86–89) and the viscera table (circles 74–81). Through the wide windows of this office, the managers have a sightline to almost every area of the kill floor's clean side. Suggestively, the killing work on the dirty side remains concealed from the manager's office behind the opaque wall dividing the clean and dirty sides. From their wide windows, the kill floor managers monitor the white, red, yellow, green, orange, and blue hard hats, using a simple color schematic to determine with one sweep of the eye whether everyone is in his or her proper place.

The slaughterhouse turns a chameleon face outward, blending seamlessly into the local urban landscape. The symptoms of its existence—the coming and going of hole-pocked trailers, the municipal signs soliciting reports about manure and odor—underscore the continual skirmish between a modern economy that requires meat without emotional responsibility for the killing of the animal and the incessant, uncompromising demands of organic life.

But if the exterior of the slaughterhouse is flat and homogeneous, fitting nondescriptly into its surroundings, it is not so with the space within those opaque walls: a highly variegated terrain with its own front, middle, and backstage spaces, its own mountains and valleys of visibility. The slaughterhouse that stretches from front office to fabrication department to cooler to kill floor makes differing demands on the people who populate it, requiring different kinds of awareness about and experiences of the work of killing to which each makes a contribution. Yet even on the kill floor itself, the site where one might least expect the realities of killing to be sequestered, immediate and visceral confrontation with the work of industrialized killing is neutralized through a division of labor that finds its sensory expression in a meticulous partitioning of space.

IV
"Es todo por hoy"

Job Number 53, Ear Cutter: *uses hand knife to cut ears completely off cow's head. Discards ears in gray barrel.*

In March 2004 I traveled to Omaha for the first time to survey the area and get a sense of the likelihood of gaining employment in one of the industrialized slaughterhouses there. At that time I made no personal contact with anyone in the slaughterhouses, limiting my interactions to taking photographs of the exteriors of the buildings, mapping their geographical locations, and making telephone inquiries about the availability of entry-level work. The response I received from my telephone inquiries was invariably some variation of "We're always hiring. Just show up in person." This indicator of the ready availability of jobs in the slaughterhouse fit what I knew statistically about the high rate of employee turnover in the industry (some estimates put the average turnover rate at more than 100 percent a year) and was further bolstered by large, permanent billboards on the main city thoroughfares near one of the slaughterhouses. These billboards carried a logo of the slaughterhouse and read in bold black print: "JOBS: Employment Opportunities in Production, Maintenance, and Sanitation. Apply in Person Today! An Equal Opportunity Employer. TRABAJO: Oportunidades de Empleo en Producción, Mantenimiento & Saneamiento."

In late June, I moved from the East Coast into a low-income Housing and Urban Development cooperative in Omaha. My partner and two daughters planned to join me after I found work.[1] As I had done during my March survey trip, I phoned several plants in the area, inquiring about the availability of work. At one slaughterhouse I was told to report to the guardhouse at the plant between seven and eight in the morning and talk to someone named Juan. At two others I was instructed to apply in person during normal business hours. No one asked me for a name or phone number.

On a Wednesday morning in June, I drive to one of the slaughterhouses, arriving a few minutes before seven. The prairie sun is already a glowing hole in the eastern horizon. I suppress a wave of nausea at the stench in the air and park at a pet-food store across the street, irrationally anxious that my Subaru station wagon will somehow single me out as a researcher. In fact, I am anxious that everything about me will give me away. My strategy of gaining access to the slaughterhouse as an employee rather than a visitor or a researcher results in a constant edginess over the risk of being "discovered," and this acts like background static—sometimes muted, sometimes deafeningly loud—every moment of my time in the slaughterhouse. I wear blue jeans, a pair of old work boots, a gray polo shirt, and contacts in place of my glasses, not so much to hide any "intellectual" look as to appear ready for the physical rigor of slaughterhouse work. I am twenty-eight, my skin is brown, and I grew up in Thailand; my hope is that the combination of my youth, sex, skin color, and non-United States background will fit the hiring manager's perception of slaughterhouse laborer material.

I cross a wide street busy with cattle trucks. Immediately in front of me is a painted plywood sign with the name of the slaughterhouse and, beneath the big block letters, two black directional arrows pointing in opposite directions. Beside the first, pointing to the right, are the words "Front Office; Sales; Visitors." The second, pointing to the left, reads, "Employment; Shipping." Even before I have placed a single foot on slaughterhouse property, I have discovered that segregation of workspace operates as a basic mechanism of power. I look toward the right, following the arrow for "Front Office; Sales; Visitors." A white concrete driveway opens up into a small parking lot surrounded by a lawn. At six in the morning, the lot is all but empty.

Only my gaze follows the arrow indicating "Front Office; Sales; Visitors." My feet follow the left arrow, "Employment." I am here to participate, not visit. Unlike the curving driveway and manicured lawn that frames the space to the right, a small rectangular trailer on raised cinder blocks immediately confronts anyone turning to the left. Behind this trailer is a chain-link fence topped with three strands of barbed wire and behind that an enormous asphalt parking lot, almost full, with a motley assortment of old and new cars and pickup trucks, some so badly rusted that entire quarters of doors seem to be missing, and others so new that they gleam in the early morning sun.

A section of the chain-link fence immediately next to the raised trailer is open, but a bar, marked with orange and white stripes in badly peeling paint, blocks the entry. I approach the raised trailer. A man and woman in dark-blue uniforms with patches reading "American Security," orange fluorescent vests, and dark glasses are sitting inside. I climb the three makeshift wooden steps to the window of the trailer and look in, attempting a friendly smile. The man, who is sitting closest to the window, looks at me and lifts his head slightly, raising his eyebrows. He slides the glass window open. "I'm looking for work," I say. The man thrusts a clipboard with an uncapped pen tied to it out the window at me. I record my name, number 26 of the day. He takes the clipboard and jabs it in the air toward a long rectangular trailer, also raised on concrete blocks, just behind the chain-link fence. "Go," he says.

As I edge around the tip of the orange-and-white entry gate, it rises suddenly, and a big truck hauling a refrigerated trailer with "Wholesome Foods" printed in large red lettering on the side rumbles by, its massive wheels just inches from my feet. Like trucks and live cattle, humans seeking work are directed to the rear of the plant. Laborers and would-be laborers are just inputs in the production process.

The trailer behind the guard hut is dull, off-white corrugated sheet metal with wooden steps leading to a closed door. It has windows spaced at regular intervals, but the shades are closed, making it impossible to see inside. Five brown-skinned black-haired women mill around the bottom of the steps. They look young, maybe sixteen to twenty, and are talking in Spanish. I nod hello and smile, but they ignore me. I walk around them, go up the trailer steps, and open the door.

It is dark inside; the only light comes from a dim bulb in the middle of the room. It takes a moment for my eyes to adjust, and when they do I see that the walls of the trailer are lined with a brown-wood veneer, much of it bulging with air pockets or peeling. Wooden benches line the long walls of the trailer, and a gray plastic fold-out table is pushed against a short wall. On a corner of the table is a yellow telephone, and under the table is a two-drawer gray metal filing cabinet with a big padlock on the bottom drawer.

The benches are filled with people. I count twenty-three, all with dark hair and eyes. Skin tone in the room varies from white to black, with most like mine, a light brown. There is a mix of ages. There are young women, dressed for the most part in jeans or sweats and tight midriff shirts that show their belly buttons, and young men, wearing baggy jeans and hooded sweatshirts. Middle-aged and older women wear dress pants and printed shirts, and several have handbags and purses. One is wearing a long-sleeved ankle-length dress. There are middle-aged and older men, wearing jeans, long- and short-sleeved dress shirts, cowboy boots, and cowboy hats. Without exception, they all have facial hair: a mustache, a goatee, a full beard.

A young woman with black hair and brown eyes is perched on the edge of the gray plastic table. I catch her eye and she smiles. "Do you need an application?" she asks in English. So much for passing, at least in appearance, for a non–English-speaking immigrant, I think to myself. I nod yes, and she opens the top drawer of the filing cabinet and pulls out a sheet of paper and a pen. "Thank you," I say, taking them.

There is not much room on the benches, so I step back outside to fill it in. The application includes a place to mark where I wish to work: slaughter, fabrication, or warehouse. I check "slaughter," then move on to name, social security number, prior work experience, and contact information for two references. Having anticipated this, I write down the names of two people who have agreed to serve as my references. Nowhere on the application am I asked about my educational background, but there is a place to note whether I have any previous experience in a slaughterhouse. No, I write, but I have worked with live cattle on a ranch, thinking of the year I intermittently helped out on a ranch in rural Oregon while an exchange student from Thailand. I print using sprawling capital letters, afraid that my writing, my spelling, my syntax—something, anything—will betray me.[2]

In those moments outside the trailer, I confront other pitfalls and limitations of participant observation: what if I cannot get a job in this slaughterhouse, or in any slaughterhouse? The trailer inside is full of people; surely they will not hire us all. I feel a familiar competitive urge—I will be one of the people hired. I have to be hired. I must be hired. My entire project depends on it. And shadowing this urge comes another doubt, tinged with guilt. What if I am hired, and this denies someone else a desperately needed job? By what right am I here competing for jobs? I feel this double-edged anxiety and guilt keenly, along with the weight of my identity, my writing style, and my idiosyncrasies, all of which I am trying to modify in my attempt to participate and observe.

I hand the completed application back to the woman inside the trailer. "You can wait outside," she says, smiling again, and I turn and step outside again. By now there are more people standing around outside the trailer in small groups or by themselves. A few are also filling out applications. Trucks come and go, idling as they wait for the bar of the parking-lot gate to go up so they can pass. Every time the wind shifts, it creates a new variation on the overall stench in the air.

After about eight more trucks pass through the gate, a scowling large-framed man wearing black jeans, leather boots, a black polo shirt with a black insulated vest over it, a white hard hat, and a radio saunters around the guard shack and walks briskly up the steps into the trailer without looking at any of us. The young women outside the trailer stop talking as soon as they see him and head up the steps behind him. The rest of us follow them and crowd inside the trailer, standing between the rows of people seated on the benches.

The man sits on the table, which sways under his weight, and exchanges a few inaudible whispers with the young woman. He takes the stack of applications from her and sifts through them, periodically shaking his head and murmuring to himself. The eyes of everyone in the trailer are on him, and necks are craned toward the applications he is looking at. He thumbs through the applications several times—there are only about seven in all, which surprises me, given that there must be at least thirty people in and around the trailer—and each time he shakes his head slowly as if dismayed. His hands are massive, his right index finger severed at the second knuckle. I surmise that this is the Juan I was told about over the phone.

As the moment stretches on, my anxiety about not being hired shades into panic. When I see my application surface once again on Juan's third sift through the pile, I do what I had seen an older man do.

"That's me," I say.

He glances up quickly at me and says in a slow deliberate voice, "We don't have anything right now. What I am looking for is someone with experience with a knife. You can check back again later and we might have something."

"I am happy to work in sanitation or fabrication or anything else you might have," I tell him, desperate. He looks at me and repeats that I should check back "later." Aside from the old man who had spoken up earlier, no one else in the room has said anything. I fall silent, my cheeks burning. I feel like an over-eager pupil raising his hand and shouting answers out of turn.

Finally, Juan tosses the applications on the table and looks around the room slowly, pointing to an older man wearing a cowboy hat. "Identification," he says. The man gets to his feet and pulls a wallet from his back pocket, handing over a driver's license and social security card. Next another older man and a younger woman sitting beside him are singled out, and they too give Juan their cards. He motions the three of them outside.

Juan studies the six cards on the table in front of him and announces without looking up, "Es todo por hoy—that's it for today." The young woman takes out a key and unlocks the padlock on the second drawer of the filing cabinet. She pulls out a stack of completed applications about a foot and a half thick and begins sorting through them, pulling out the three that match the identification cards on the table. The rest of us file slowly and soundlessly out of the trailer. The three new hires sit across the parking lot on the edge of the curb, avoiding eye contact with us. One younger man in front of me says suddenly, to no one in particular, "Oh come on! I can't believe they're hiring all these old people."

Later that morning, I drive to another industrialized slaughterhouse located near the first one. In contrast to the temporary trailer at the first slaughterhouse, the employment office at the second is housed in a brick office building across from one of the four city streets that border the main plant and is open every day from nine A.M. to three P.M. When I step into the building, there is only one other person inside, a tall, thin black man filling out an application at one of three long plastic tables. He looks up at me as I enter and we exchange greetings. The room is well-lit and air-conditioned, with a soda machine in the corner and a calendar hanging on one of the white walls. A paper sign reading, "DO NOT KNOCK OR ENTER IF DOOR IS CLOSED: INTERVIEW IN PROGRESS," is taped to a half-open door at the end of the room.

I knock on it and a male voice says, "Yeah." Seated behind an L-shaped table is a lean, thirtyish white man with a red mustache. He looks at me expectantly, and I tell him I'm looking for a job. He says, "O.K., fill this out," and hands me several pieces of paper, motioning me back to the other room. The black man is still sitting there, thumbing through a small appointment book and copying things from it onto his application. I sit at one of the other tables and fill out the application, which is similar to the one at the first slaughterhouse.

When I am finished, I walk back into the office, where the red-mustached man now has his feet up on the table and is eating Ramen noodles out of a Styrofoam cup. He takes my application and throws it onto a pile of papers without a glance. "We're full right now but be here on Monday at 6:15 in the morning."

"What are the chances of something being open then?" I press.

His response is irritated and deliberate: "Look, I said we're full right now, but I wouldn't tell you to come back on Monday if I didn't think something was going to be open."

I thank him and make a quick exit, waving good-bye to the black man still filling out his application.

The next morning, I return to the first slaughterhouse, arriving at quarter to seven. The rough outlines of a certain scripted ritual quickly make themselves clear. There is little to do but show up, sign my name at the guard hut, take up a position with the other applicants in the trailer, and hope that Juan's severed finger will single me out before the words "Es todo por hoy" once again drive us back into the world of the unemployed. There are already sixteen people inside the trailer, and I see many familiar faces from yesterday. Aside from those who have come together, there is no conversation, just a continuation of yesterday's tense silence. A handful of newcomers fill out applications. Their faces betray the most emotion, the most hope. For the rest of us, the day of application has already passed, the line into supplication already been crossed.

The extreme anxiety, the feeling that getting in is close to impossible, the dramatic transformation from applicant to supplicant enacted each morning when Juan makes his appearance furnishes a powerful reference point for all of us who will eventually be hired to work in the plant. "If you don't like it here, there's the door," we will be told repeatedly. We understand from our experience in the employment trailer that the door is likely to be a one-way exit. For those limited by language or citizenship status to all but a few work opportunities, the scripted ritual of the employment trailer works as hard at keeping them locked in the slaughterhouse as it once did at keeping them locked out of it.

Juan enters the trailer at five after seven. "Buenos días," he says. "Buenos días," we murmur in unison. He takes the new applications from the young woman. One in particular seems to interest him. It belongs to a stocky brown-skinned man wearing a cream-colored shirt and white pants. He asks the man in Spanish what experience he has and the man replies, "Incisor." Juan nods and asks to see the man's identification. Juan takes his driver's license and social security card, examines them, and asks to see a legal residence card. The man shakes his head. Juan shakes his head back; no one can work here without a legal residence card.

As we sit quietly, Juan speaks into his radio a couple of times, asking someone named Jason, "What is the count right now?" The reply comes back over the radio, "Two seven six. Two seven six." Juan nods in apparent approval. After shuffling through the applications a couple more times, Juan abruptly announces, "Es todo por hoy," then, looking at me, adds in English, "That's all for today." Half the group start moving for the trailer door; the other half, including the group of young women I had seen outside the trailer yesterday, stay seated as if to say, "No, that is not all for today. We want jobs." Juan repeats, louder, "Es todo por hoy!" This time, no one remains seated.

As we file out the trailer door, I linger and turn to Juan. "Is every department full, even slaughter and sanitation?" I already know the answer, but I want him to know how badly I want the work. I also want to feel out the human being behind the role. He is surprised, but not offended, by my boldness. In deliberate, accented English he says yes, what he needs are a couple of people with experience chuck boning. He says he just cannot hire anyone without experience at the moment. I tell him that I am a fast learner and a hard worker. He smiles at this earnestness and says, not unkindly, that I will just have to come back and check next week. Bantering now, I tell him I will be back tomorrow, just so he won't forget my face. He replies, smiling, "Oh don't worry, I won't forget your face. I have a photographic memory, I remember everyone's face. Some guys worked here two or three years ago and try to come back and think I won't remember who they are but I do."

At this, the woman who hands out applications nods and says, "It's true, he even remembers their names." Emboldened by this brief exchange, I press on. "What are your names?" I ask. He is Juan, as I guessed. The young woman's name is Michelle.

As I reach my car, rain starts to fall, and a clean smell competes with the stench in the air. A man with a black mustache in blue jeans and a black T-shirt walks in front of me. He turns to cross a pedestrian bridge that spans the interstate, walking straight into the drenching wetness. That morning, he has neither spoken to nor been spoken to by anyone. Just "Es todo por hoy."

Fourteen people are in the trailer on Friday morning at seven. Five young men and two young women are filling out applications, meaning there are seven of us back again from previous days. Four of the new young men have come as a group, and one is serving as a coach for the others, supplying them with various numbers needed on the application. One of the new applicants is a black man, probably in his thirties. He wears a blue shirt and checks a cell phone compulsively.

Michelle comes in at ten minutes after seven and says, "Buenos días." The phone rings, and she answers it, speaking for about five minutes. I hear her say, "How many people? Two? Three?" After she hangs up, I say, "Good morning, Michelle." She says hello and tells me that they are looking for a couple of people over in slaughter, no prior experience necessary. She says that hopefully I will get hired today, but only if Juan comes out. "He usually doesn't come out on Friday," she adds.

Five minutes later, Juan enters the trailer. He goes through the new applications and looks up at the African American man. "You are missing the last two months here," he says. The man tells him he was working as a security guard. Juan says, "Come back on Monday and we'll see what we can do for you." "Monday?" repeats the man. Juan nods, and the man leaves the trailer.

After that, Juan looks around the room and makes a general announcement in Spanish that there is nothing available at the moment but people should check back in the next week or two. One man asks whether they will get a call from the plant if something opens up. Juan says no, they will have to come in person and check for themselves, then he nods at me to stay.

Everyone except me and one other man, who is still filling out his application, leaves the trailer. Juan asks Michelle to get my application out of the filing cabinet and asks me for my identification and social security card. I am elated, the winner of a prestigious prize. From the multitudes, I have been handpicked for the privilege of slaughterhouse work.

While Juan looks over my paperwork, I question him about what I will be doing (hanging livers in the cooler), working hours (six in the morning until whenever the work is done, usually about five or so), working days (six days a week, but they haven't been working on Saturdays for the past four weeks), equipment (they will provide me with boots and gloves and a hard hat, but I need to wear warm clothing because I will be working in the cooler), and pay (he doesn't know; I'll have to ask my supervisor). I ask him what kind of cattle they process (Angus and Hereford), how many they slaughter a day (about twenty-five hundred, roughly three hundred an hour), and how many people work here (more than eight hundred). I make a joke about being glad that I'm hanging livers and not ears since that means only twenty-five hundred livers a day instead of five thousand ears. Juan laughs and says, "Just be glad you aren't doing feet because that would be ten thousand a day." I ask where the livers are going, and he says they are for export to Russia and the Far East; the slaughterhouse just landed a contract for liver exports, which is why they need some new people.

Juan writes my driver's license number and "Soc. Security OK" on the top of my application. He then sends me outside with Michelle to head over to the slaughter side of the plant. He is, Michelle tells me, in charge of the fabrication side of the plant, where the cattle go to be cut into smaller pieces after they have been killed, and he also does all the hiring. She tells me that he is the highest-ranked Latino in the plant and that he is not necessarily liked by the employees since he can be "hard."

Michelle asks me whether I have a car and when I say yes she tells me to get it and park it in the lot. I drive up to a second gate in the chain-link fence marked "Employees," and the guards open the gate for me. I meet Michelle by the employment trailer and follow her past a row of refrigerated semitrailers backed up against a concrete loading dock. We are walking around the perimeter of the large concrete structure that looms over the front office. At the side of the structure, near the docks for the refrigerated trucks, the building material changes from concrete to a bright white aluminum or metal siding, and the number of turbine-like fans on the top of the flat roof increases.

"Have you ever worked in a meatpacking plant before?" Michelle asks me as we walk together. "You're in for a big surprise," she continues when I shake my head. "It's different. It's real different." I press for details. "Some people just can't handle it," she says. "We hired this one guy, and after two hours he asked to go to the bathroom and then never comes back. Some people stay for a day. Some people stay for a week. On the fabrication side, it's not so bad. The blood in the meat has already been frozen. Slaughter, though, that's bloody and dirty."

"It just depends on if you like your job or not," she adds after a pause. I confirm that I will be working on the slaughter side. She says, "Yeah, but you'll be in the cooler so it won't be so bad there. But you better dress warm," she warns.

Michelle seems friendly enough, so I ask her the question that has been bothering me: "Why did Juan hire me?" She shrugs her shoulders and laughs. "I don't know, I guess he likes you because you're nice."

Continuing around the perimeter of the building, we come to a set of double glass doors. Set into the wall above the doors is a large sign with the slaughterhouse logo on it reading, "Through these doors walk the finest food production team in the world." This is the employee entrance. It lies on the opposite side of the building from the front office entrance, away from the main city road, and on the same side where the cattle are unloaded to be slaughtered. Michelle holds the door for me, and I walk into a long, narrow corridor. The two-story walls are of rough concrete. Several people sit on low, narrow wire-mesh benches, smoking cigarettes as they lean back against the walls. They have on white hard hats and are dressed in heavy clothing, even though it is warm in the corridor. The smell is hot, damp, and suffocating.

Halfway down the corridor against one of the walls I see a glass-encased bulletin board with Occupational Safety and Health Administration (OSHA), minimum-wage, and sexual-harassment policy notices in tiny print; they are all but invisible behind the glare of the plastic laminate. At the end of the corridor is another set of glass double doors, identical to the exterior ones. Passing through these, I see a third set of double doors immediately in front of me. To the left is a long, wide hallway with a polished concrete floor and white-painted cinder-block walls. To the right is another wall, this one with a brightly lit board reading, "Communication Center," with two identical posters—one in Spanish and one in English—titled "Got Resentment?" and, in large black print superimposed on a humorous picture of a scowling man, the injunction "Resentment Can Poison Workplace Relations, Increase Stress, and Lower Productivity; Take a Deep Breath; Count to Ten; Is This Issue Worth Blowing Up Over? Try to Understand Where the Other Person Is Coming From." Next to the poster is a flyer announcing the upcoming annual company picnic.

In some of my lowest moments during the months ahead, this Communication Center will offer an unfailing source of bitter humor. Nothing of true importance, such as whether we will be required to work on an upcoming Saturday, is ever posted there. Instead, there is an ever-changing array of glossy inspirational posters. One shows a picture of a tree with bright orange leaves with the words,

> Losers make
> Promises they
> Often break.
> Winners make
> Commitments they
> Always keep.
> —Dennis Waitley, motivational speaker and author

Another is titled "Worth Thinking About: A Word of Thanks, Please," followed by an anecdote about a worker who stayed up all night completing a project for his boss only to have the boss snatch up the work the next morning without so much as a word of thanks or gesture of gratitude. The miserable story ended with the exhortation "Others may be crass, but let us show class!" Still another poster is titled "Illustrated Current News: Things We Don't Quite Understand" (International Edition). This quotes a study by Yale University's Child Health Research Center about why people stop growing when they do.

These posters strike me as sadly comical. Mass produced by a company in North Haven, Connecticut, they seem to be targeted for a different working regime, perhaps an office or a bank. They give the message—glaringly contradicted by almost every aspect of our everyday experiences—that we, the employees, are human beings with real emotions (albeit emotions to be managed in the interest of productivity), a need for respect and appreciation (though expected to show "class" even when that appreciation is denied by the "crass" behavior of superiors), a sense of intellectual curiosity (to be roused by such problems as why people stop growing when they do), and an interest in personal character development (under the guidance of inspirational speakers and authors). Further, because this conspicuous Communication Center is duplicated in other places by different message boards—these without fancy backlighting or the need to announce themselves—that carry information that every worker really does stop to read (such as whether an upcoming Saturday will be a workday), I sense that the self-proclaimed Communication Center is a pet project of the front office, a fiction encroaching on the outer reaches of the production plant.

I follow Michelle down the hallway to the left, the soles of my boots slipping on the slick concrete. In the middle of the hallway to the right is a set of stairs as wide as a car is long. We climb the concrete steps, make a 180-degree turn at a white cinder-block wall, then climb the rest of the way to the second floor and another long hallway of polished concrete. At the top step, Michelle calls, "Ricardo," to a short stocky man in a dark-blue uniform and a red hard hat who is walking by. The man has light-brown skin and small dark eyes set into a round pudgy face. His lips hide under a wispy mustache that seems to grow straight out of a prominent mole on the side of his nose. Michelle introduces me to Ricardo as "the new guy for the cooler," and he tells her, "Just do his orientation today." Looking at me briefly, he says, "Be here on Monday at 6:30."

"Where should I go?" I ask.

"Just stand right here," he answers in a slightly accented voice. "I'll meet you right here."

"He's worked with cows on a ranch before," Michelle says suddenly.

Ricardo looks up at her and says, "Oh."

"Will there be a chance for me to do things other than in the cooler?" I ask.

Behind me, a voice says, "Oh yes, there is plenty of opportunity here for someone who works hard, shows up at work every day. You can go right to the top, as Michelle knows. Michelle started out here folding boxes and now she's processing new hires."

I turn my head and see a tall slim white man with dark eyes, a prominent nose, curly dark hair sticking out beneath a white hard hat, and a thick mustache. He is wearing a dark-blue overcoat of some thin material, opened to reveal a light-blue polo shirt, blue jeans, and dark work boots. He extends his right arm toward me, and I turn awkwardly and shake his hand. "Bill Sloan," he says, before turning and walking down the hallway, followed by Ricardo.

Ricardo, I later learn, is one of many supervisors in the plant, but unlike other supervisors, who are assigned to manage an actual production area, Ricardo functions as a kind of “roaming whip” for the plant management, acting as a translator in disciplinary situations and as a job broker when different positions become open for internal advancement. Bill is the son of the kill floor manager, officially second-in-command on the slaughter side of the plant, and has authority over almost all personnel decisions and much of the day-to-day decision making in the plant. Although he is deferential whenever his father is involved, the broad scope of his authority lies in the way in which situations are or are not brought to his father’s attention. There are line workers who hold no official title in the plant but because of their close relationship with Bill can get away with doing little or no work and with flouting the authority of their immediate supervisors.

Michelle takes me back down the stairs past the brightly lit Communication Center. We turn left through the third set of glass doors and immediately left again into a large open room ordered by rows of picnic benches and filled with the low buzz of conversation. Outside the room, a sign reads, “Attention Slaughter Employees: This Is a Restricted Area: Permit Needed for Entry.” Hundreds of people in white overcoats and white hard hats sit together at the picnic benches putting food into their mouths at a furious pace. Given the summer heat outside, they look like transplants from another climate, bundled in layers of sweaters and white frocks. A few look up at Michelle and me as we walk past, and I smile and nod my head at them, receiving a few smiles in return.

Halfway down the room Michelle turns and opens a door with no markings or signs on it. We enter a small rectangular room with the same white cinder-block walls I have seen everywhere else. There is a plastic table in the middle of the room and another with a small television on it pushed up against the far wall. A glass showcase sits against the third wall. Inside are various trophies, laminated pictures of soccer teams, and plaques from the American Meat Institute commending the company for its superb workplace environment and excellent safety record.

Michelle tells me to sit in one of the metal folding chairs and says I will be watching a video. She then hands me a folder with information about working at the plant and my benefits. The opening credits identify it as a safety video produced in Omaha for the Red Meat Industry with a grant from OSHA. What follows is an hour of mind-numbing do's and don'ts (do tell your supervisor if you see any conditions that are unsafe, do wear your hard hat at all times, do stretch and exercise at home; don't run, don't lift with your back, don't operate a forklift unless you are trained to do so, and so on) and warnings about various injuries "common to this type of work" (crushing, cutting, repetitive-motion disorders, back injuries, chemical burns, and the severing of fingers, hands, and other body parts) accompanied by images of injury-causing scenarios which always stop short of showing any actual injury (a forklift careens wildly out of control and heads dramatically toward a worker, but we never see it hit the worker; a worker runs with a knife and appears to slip, but we never see him cut himself).

I try to concentrate on the information in the video, but the fast, monotonous narration and the repetitive images remind me of the rapid-fire legal disclosures that are added on to the end of radio advertisements for car dealerships and mortgage financing. Not only that, I am still so elated at being hired that I cannot focus on much of anything. After fifteen minutes, I give up trying to memorize the list of do's and don'ts. As I will discover after a few weeks in the plant, the list is almost the exact inverse of what actually happens on the production line (never tell your supervisor about an unsafe condition, do run, do lift with your back, do operate a forklift whether you know how to or not). After thirty minutes of the video, it seems as though there is so much danger in working here that I won't even make it through my first week alive. After forty minutes, I sense that if anything—anything at all—happens to me while I am working in this plant, the slaughterhouse will be able to say I was warned about it during my orientation video. Michelle walks in and out of the room at intervals, smiling and asking whether I have any questions about what I am watching.

"Did you understand everything?" she says when the video is finally over.

"I think so," I reply, hoping there won't be a quiz.

There is not, but there are a multitude of forms. There are forms stating that I have watched the safety training video. There are forms stating that I understand that the company operates on a hire-at-will basis, meaning that I can quit or they can fire me at any time, for any reason, with or without notice. There are forms stating that I agree to reimburse the company for any unreturned equipment in the event that my employment is terminated. There are forms verifying that I am legally available for employment in the United States. There are forms asking me how many deductions I want taken for my income tax. This last reminds me that I have no idea what I will be paid. I ask Michelle, who has no idea either. "Ask your supervisor," she tells me.

Next she stands me against the white wall so she can take a Polaroid of me for my I.D. card. We chat while the picture develops. She is twenty and started working in the plant two years ago. She began in the basement folding boxes, until the "other lady" who helped the front office do hiring and orientations quit, and she was promoted because she speaks good English and Spanish. She has two young children and lives in South Omaha, in a predominantly Mexican neighborhood with only "two white guys." I ask her whether this is a good place to work: does she like working here, do people in general like working here? She shrugs noncommittally: "It just depends on if you like your job or not." Michelle gives me a temporary I.D. card and tells me I will need it to get past Security on Monday. We then walk back through the cafeteria, empty now with only some colored lunch boxes sitting on the picnic tables, and through the hallway. There we see another man in a red hard hat. He is thin and wiry, with white skin and pale-blue eyes. Michelle introduces him as James. "He'll be your supervisor," she tells me and leaves me with him, walking away with a wave of her hand.

James tells me that work starts at seven and that they have been working nine-hour days Monday through Saturday. He says I will want to dress warmly because it is cold in the cooler and that I can bring a lunch with me or buy it from one of the food trucks outside. I ask him how much I will be paid and he says, "I don't know. I can check on that for you." He tells me to follow him back up the stairs to the second floor, where he leads me into a room filled with metal utility shelves stacked with boxes, paper towels, and soap. Enormous laundry machines sit at the back of the room; in one corner is a mesh cage, and an older black man with a vast, protruding stomach, gold-rimmed glasses, and a sun-spotted face sits on a black metal stool inside the cage biting on a toothpick. He is wearing a blue T-shirt and pants, leather boots, and a white hard hat. A black radio is clipped to his belt.

"This is a new guy," James tells him. "He's going to work in the cooler on Monday. He's going to need boots, gloves, a white frock, and a white hard hat." The man nods and says, "Okay."

James tells me to come back to this room on Monday about fifteen or twenty minutes before seven. I thank him and walk down the stairs, out the double doors, through the hallway, out another set of double doors, and into the bright glare of the noon sun.

Of his fictional city Esmeralda, Italo Calvino noted: "A map of Esmeralda should include, marked in different colored inks, all these routes, solid and liquid, evident and hidden."[3] As with Esmeralda so with the physical and relational spaces of the slaughterhouse. It would take months before I was able to mark the solid, evident routes and longer still before I could draw, in a different colored ink, the liquid, hidden ones. On this, my first day, each workspace, each passageway, each door and wall, was a strange, alien thing and each face, each voice, each outstretched hand an unknown. There were clues—the manner of dress, the color of the hard hat, the authority of the voice—but these, as I would come to discover, misled as often as they informed.

V
One Hundred Thousand Livers

Job Number 119, Supply-Room Staff: *controls distribution of equipment to workers, including hairnets, gloves, safety gloves, aprons, knives, rubber bands, plastic bags, and hard hats.*

"Hello, asshole."

The words float back from the front of the line, low and unmistakable. It is 6:45 A.M. on my first full day as a slaughterhouse employee, and I have had the weekend to make some sense of the confusing jumble of safety and training videos I was shown on Friday and to mull over what it might mean to be a "liver packer." Anticipating the cold working environment, I have on long underwear under my jeans and two layers of sweatshirts on top of short- and long-sleeved T-shirts. I also have a small lock and key, which I was told to bring for my locker, and a sandwich with a piece of fruit in a plastic bag.

"Motherfucker."

It is the same raspy voice, barely audible, drifting back over the column of white and gray hard hats. Up on my toes, peering ahead, I see the black man James introduced to me on Friday sitting on a stool just inside the cage. His feet rest on the bottom rung of the stool while his hands shuttle back and forth to a supply shelf, dispensing white cotton gloves, green rubber gloves, black hairnets, and little plastic packages with small orange cones—earplugs—inside. The men in the line silently move forward to take their turn in front of the cage.

"Good-bye, asshole."

Closer to the front now, I see the lips move and know with certainty that it is the black man speaking these words, these curses that accompany the gloves and the hairnets. The man in front of me, shielded by the man in front of him, raises his middle finger in concealed response. Then it is my turn at the front of the line.

"You're new," the black man declares. "Where to?"

"The cooler," I say, "to hang livers."

"You won't need a knife then. What size shoe do you wear?"

"Nine and a half."

Agile for his bulk, the man slides off his stool, walks to the back of the cage, and returns with a gleaming white hard hat, a pair of green rubber boots still wrapped in plastic, and a white coat. He adds a pair of white cotton gloves, a pair of green rubber gloves, a black hairnet, and a package of earplugs. "Tomorrow morning, bring the white gloves back to get new ones," he says. "And sign this," he adds, thrusting a clipboard at me. I glance quickly at the writing on the paper, aware that I am holding up the line. The form states that I have received a tour of the plant and training in emergency exit procedures, am aware of the location of first-aid stations, and acknowledge that the costs of any equipment not returned "in clean and good condition" will be deducted from my final paycheck. I ask about the tour of the plant, which I am anxious to have.

"Oh, that, well, I can't really show you around because I don't know where anything is anyway."

I take the hint and sign my name, then move to the side of the room with my armload of equipment. It will be several weeks before I come to know this black man as Oscar, supply-room manager, thirty-year veteran of the industrialized slaughterhouse (with a paid cruise from the company as a reward for his "loyalty and dedicated service"), afternoon school-bus driver, and casino aficionado. Often so exhausted that he falls asleep while eating during break, Oscar alternates between telling the slaughterhouse workers, "You are a good man," and running through the invectives already mentioned. The only consistent pattern I can determine is that the workers who share their food with him are "good men," while those who mock him for falling asleep fall in the other category.

A six-foot-plus white man, so thin that he is almost gaunt, walks into the supply room. He wears thick-rimmed glasses, a light-blue short-sleeved polo shirt tucked into dark-blue dress pants, and no hard hat. Spotting me, he asks, so softly that I have to lean forward to hear him, "Are you new? Do you know where the locker room is?" When I shake my head he motions for me to follow him out of the supply room. Around a corner at the end of the corridor, he gestures at a door with a sign reading, "Clean Men's Bathroom."

"The lockers are in there," he says; "you can leave your stuff in there, and you need to bring a lock for your locker."

"Thanks," I reply. "What's your name?"

"Me?" he asks, surprised, then adds, "Rick. I'm Rick," before walking back down the hallway and turning the corner. As with Oscar, it will be several months before I learn more about Rick. Hired as a "safety coordinator" for the kill floor side of the plant, he deals with OSHA record keeping and requirements, administers first aid, coordinates the annual health checkup for the employees, and helps the food-safety coordinator in the front office perform tests for microbacterial levels at different locations in the plant. A soft-spoken evangelical Lutheran who invites me on more than one occasion to attend his church, Rick is the son of Iowa corn and soybean farmers. He completed a degree in engineering at Iowa State five years ago, and his first and only job since graduation has been at the slaughterhouse. Almost universally ignored by workers, supervisors, and managers alike, he is rarely on the kill floor; most of his time is spent in his small, windowless "Safety Coordinator Office" at the end of the hallway.

The clean men's bathroom has five stalls, four urinals, and four sinks. At the other end of the room, a door with a sign reading, "Clean Men's Locker Room," opens into a dingy, malodorous room with three parallel rows of gray metal lockers, one against each of the two walls and a double-sided row down the middle, with narrow white benches between them. A small hallway running between the bathroom door and the locker room opens into a tiled shower room with five showerheads placed at regular intervals around the walls.

I walk down the first row of lockers, nodding at several men in various states of undress, find an empty locker, number 225, and sit on the bench to figure out my equipment. The hairnet is straightforward: stretching it between two hands I slip it over my head. The hard hat consists of a tough outer plastic shell lined with gray plastic strips inside. It is these strips that make actual contact with the top of my head, leaving something of an air cushion between my head and the outer shell. The plastic band that adjusts for width takes some fiddling with. There is a comfortable girth, but that does not hold the hat on my head if I lean over or move too quickly. The next notch leaves me with a headache by the end of the day but holds the hard hat on during any kind of motion. The green rubber boots are calf high and hard to get on, but feel snug and surprisingly comfortable. I open the small plastic package of earplugs, which consist of two orange foam cones connected by a strip of blue plastic. I try to push the cones into my ears, but they slip back out; resting loosely in my outer ears, the plugs do not do much to block out noise. Not sure what to do with the two pairs of gloves, I stuff them into the pockets of my jeans. As the locker room empties out, I hurriedly put the rest of my belongings into my locker, lock it, and head for the supply room.

Ricardo, the red-hat supervisor I met the day I was hired, finds me in the hallway and says something into the radio. Within minutes, a wiry man about five feet tall in a dark-blue uniform joins us. He has a mustache and sports a ponytail under a bright yellow hard hat. Some of the color in the dark brown iris of his left eye seems to leak into the white around it, a disconcerting effect more than compensated for by the broad, welcoming smile he beams at me, the first I have seen all morning. He looks me over and without comment walks into the supply room, returning with another white frock, which he hands to me saying, "It's going to be cold." His voice is fresh and young, inflected with the accent of a non-native English speaker who has almost, but not quite, succeeded in imitating a native speaker.

I put the second white frock on over the first. The hallway is at normal room temperature, and I feel sweaty and constricted in my multiple layers. The young man moves quickly out of the room and down the hallway. Instead of making the turn toward the clean men's bathroom and locker room, he goes through a set of large double doors.

We are in a cavernous room of gleaming, clanking metal. Directly in front of us, a metal conveyor belt about five feet wide and twenty or thirty feet long churns toward us, the bright overhead lights glinting off it. Everywhere I look there are workers in white hard hats, both on the red floor and up on elevated platforms. They are standing strangely still. Some have their backs to us, some are facing us. It is confusing; I cannot discern a pattern. The smell is an odd mixture of organic and chemical, the way a dirty bathroom might smell after it has almost, but not quite, been successfully cleaned.

It is even warmer in here than in the hallway. We move quickly through the room toward a wall with bright lights, hooks, and the same strange distribution of clumps and lines of motionless workers. We make a sharp right turn and suddenly find ourselves at the top of a wide, dimly lit flight of concrete steps about fifteen feet across that descends sharply to an enormous dull-white door, at least fifteen feet high and twelve feet across. Here the olfactory skirmish between organic and chemical has shifted strongly in favor of organic.

Gripping a metal handrail, the man in the yellow hat trots down the steps, the sound of his boots echoing off the walls. At the bottom he yanks open the door by a small metal handle.

A cloud of steam rushes out, caused by the violent collision between the near-freezing air behind the door and the hot, humid air in the stairwell. Shrouded and ghostly in the fog, the yellow hat beckons me into the vast chambers of the cooler. Row after row of headless, hoofless, hideless cattle, split in half and suspended by their hind hocks from overhead hooks, fill the room. Inanimate under the white halogen glow, they seem unreal, like giant plastic cow parts ready for assembly in the romper room of gigantic children: line up the rib bones on part A against the rib bones on part B and snap together to form a hollow, gutless outer shell. To complete, add head and hooves, and paint to the desired hide pattern.

Two black-mustached men, comically mismatched in height, exchange brief nods with the yellow hat and walk by us. They push open the enormous door and chain it against the stairwell wall. Their faces blur as fog fills the space, and they become moving blobs of light, visible only by the bright yellow rain gear they wear. The yellow hat has disappeared. I make him out moments later emerging from behind the first row of white carcasses pulling a barbed silver-colored cart, which he parks in front of me.

Welded from stainless steel, the cart stands about six feet high and has a solid base about three feet by six, with four wheels, two larger ones about six inches in diameter in the center width of the base and two smaller ones about three inches in diameter at the center length. At any one time the cart rests on the two larger wheels and only one of the smaller wheels; it is always angled down toward whichever of the two smaller wheels is making contact with the floor. Welded to the top of the base are two vertical steel plates about four inches wide and half an inch thick, joined across the length of the cart by pairs of steel crossbars of slightly smaller widths and thicknesses. These crossbars taper, pyramid-like, from the bottom, where they are farthest apart, to the top. Between the lowest pair of crossbars, also connecting one vertical column to the other, another crossbar is welded. On each individual crossbar and on both sides of the center crossbar are ten protruding hooks, each about four inches long and an inch in diameter at their base, thinning and tapering to a sharp point as they angle outward and upward. Ten hooks on each of the four pairs of crossbars plus another ten on each side of the center bar equals one hundred hooks in all; the overall impression is unmistakably that of a bristling steel porcupine.

While my eyes register these novelties, the rest of my body is assaulted at a more basic level. The cold air seeps through my multiple layers of clothing and within minutes settles under my skin and crawls into my bones. The mercury in the little thermometer mounted on the far side of the cooler wall hovers between 32 and 36. Overhead, enormous cooling fans produce a constant rumbling whine, sending vibrations into the walls and up through the concrete floor. I push my orange foam earplugs in deeper and for a moment they reduce the rumbling whine to something tolerable before they slip out again.

The yellow hat holds up two gloved hands, which I stare at stupidly before realizing he is telling me to put on my gloves. Fingers already numb from the cold, I reach clumsily into my front pockets and pull out the two pairs of gloves given to me in the supply room. The yellow hat has the green rubber gloves on so I start to put my left hand into a rubber glove. He catches my arm with his hand and shakes his head, then folds one of his rubber gloves down to reveal a white cotton glove underneath. It takes me forever to pull on the cotton gloves, then the rubber gloves. I sense the yellow hat's eyes watching, appraising, and my cheeks burn with a feeling of bumbling incompetence.

When I am finally gloved, the yellow hat points toward the cooler wall, and for the first time I notice the chain. Made of thick dark woven steel cable, it stretches tautly between two sprockets about a foot and a half in diameter suspended by long metal rods from the ceiling. One, just inside the entrance to the cooler, is about twelve feet from the floor. The other, about five feet from the floor, hangs near a concrete support column about ten feet from the cooler door. The thick cable stretches from the high sprocket to the low one, wraps around the low one, then runs back up to the high sprocket, producing parallel lines about a foot and a half apart. My gaze follows the parallel cables up the side of the stairwell until they disappear through a small rectangular opening at the top. All along the cable hang sharp-tipped gleaming metal hooks, spaced about a foot apart.

Without warning, the sprockets begin to turn, and the cable starts to move. The yellow hat puts two fingers up to his eyes, then points at the moving hooks. I nod my head and turn to look at the hooks. He walks to a hose coiled on the floor near the wall, fills a white bucket with water, and brings it back to the moving cable line, then takes a rag, dips it in the bucket, and wipes the hooks clean as they move around the lower sprocket. Now he motions for me to take the rag. Not a word has passed between us, but the communication has been crisp and efficient. I take the rag and wipe down each gleaming hook as it passes by. After I have done a couple of hooks, the yellow hat takes my hand, plunges the rag into the bucket, and motions for me to wring it. Then he stands by and watches. Five hooks, dip the rag, wring it. Five hooks, dip the rag, wring it. The yellow hat nods slightly, and I allow myself a small moment of exhilaration. After a weekend filled with dread and anxiety that I will fail at whatever being a "liver hanger" requires, I seem to be up to the task of five hooks, dip the rag, wring it.

Five hooks, dip the rag, wring it. Then, through the steam and fog, the first of the more than one hundred thousand livers that I will work with in my two and a half months as a liver hanger rounds the corner of the top sprocket. Impaled on one of the eight-inch hooks and quivering with the vibration of the cable, the reddish-brown kidney bean–shaped liver seems alive as it descends from the higher to the lower sprocket. As it moves around the lower sprocket away from the wall and toward us, the yellow hat motions for me to step aside, and grasping the liver between his two green-gloved hands he pulls it up and out. After a brief struggle he jerks it clear of the sharp hook. Dark purplish blood, at first a trickle and then a stream, spurts from the puncture made by the hook and runs down his rubber gloves, spreading slowly through the cuffs of his white frock. The yellow hat turns around, moving his hands so the length rather than the width of the liver faces the steel cart, and with deliberation half-thrusts, half-pushes the liver onto the lowermost left hook of the silver cart. Thus re-impaled, it emits another stream of blood, and the bright silver of the cart darkens with a widening circle of viscous purplish stain.

I gape at the smoothness of the liver, lined on one edge and again lengthwise through the middle with a white sinewy strip; the outline is that of an elephant's ear, but the way it curves into narrow delicate edges summons the image of a rose petal. As long as my forearm and almost two hand lengths across at its widest point, the liver is studded along the edges with fatty tunnels and openings, one of which—the posterior vena cava—the yellow hat has used as an entryway for the cart hook.

This, then, is the liver, the largest gland in the bovine

body, essential to life and possessing remarkable powers of regeneration. (A rat liver, according to Claude Pavaux's *Color Atlas of Bovine Visceral Anatomy*, will completely rebuild itself in fewer than three weeks following a partial hepatectomy that removes two-thirds of the organ.)[1] Weighing up to twenty-five pounds in the adult bovine, a quarter of which is blood weight, the liver performs a multitude of functions, including the conversion of nitrogenous waste to substances that can be excreted through the kidneys, the production and secretion of bile, the storage of vitamins and iron, and, because of its size and position in the body cavity, the production of heat to maintain body temperature. Eviscerated, isolated, and dangling on the lowermost left hook of the steel cart, this liver now radiates its steamy odor into the vast, unwarmable coldness of the cooler.

As the yellow hat motions at me to wipe the bloodied cable hook as it steadily ascends on its return path to the higher sprocket, huge, ghostly white half-sides of eviscerated and skinned cattle clank down a chain above the concrete stairway and whiz through the open cooler door, attended on either side by the tall and short mustached men in the bright yellow rain jackets. I reach for the stained hook, but it is too late—and in that reach I suddenly understand what it is I have become part of: this steamy liver, those carcass shells were just moments ago inseparable components of the same whole, a breathing, walking animal. Now, shuttled from the upper level down to the cooler via chains and cables, segregated and categorized as distinct products, they are deconstruction fleshed.

But there is neither time nor space for mental exercises in the liver-hanging department. Both livers and cattle carcasses are coming at me now, their fast pace compelling my response: action and reaction. It is surreal, this descending line of reddish-brown livers framed against the dirty-white backdrop of the cooler wall, this steady drip of blood to the concrete floor. The yellow hat works in a rhythm, pivoting on his feet between the moving cable and the steel cart, unimpaling a steaming liver, rotating it in his hands to expose the posterior vena cava, then re-impaling it on the cart. Unimpale, rotate, impale. His cadence is established, and for the next half hour we work beside each other in silence as I struggle to find a rhythm of my own. Clear the bloodied hooks of flesh and fluid, dip the rag, wring it. Clear the bloodied hooks of flesh and fluid, dip the rag, wring it. Even through the combination of rubber and cotton gloves, the frigid wetness of the wipe rag numbs my fingers. My fist clenches around the rag as I wrap it around the moving metal hooks, rubbing up and down on the stem and then back and forth on the angled hook, scrubbing off the blood, knocking to the floor bits of fat and flesh that catch in the angle of the hook.

When all the hooks on the first steel cart have a liver hanging on them, the yellow hat leans into the cart and pushes it seven or eight feet away, returning with a second, empty cart. Unimpale, rotate, re-impale; wipe, dip, wring: we have our rhythm now, this yellow hat and I, and I venture an opening.

"My name is Tim!" I shout into the two-second interval it takes for the next liver to round the lower sprocket. The yellow hat shrugs and holds his bloodied gloved hand to his ear. The overhead cooling fans are deafening. I scream it again. This time the man smiles. Tapping his frock he replies, "Javier!" After several more shouts and hand gestures I discover that Javier is from Mexico, has been working at the slaughterhouse for almost thirteen years, and considers the job just "okay" but "good money." As we exchange shouts the second cart fills up, and when Javier returns with the third, I point to myself and then at a liver. Javier raises his eyebrows but takes the rag from my hand and motions for me to take his place.

The livers are simultaneously firm and squishy. Removing them, I quickly discover that I have to apply just the right amount of pressure as I lift to clear the liver of the moving hook. Too little pressure and the liver slips. Too much pressure and it starts to collapse. Most come off cleanly, but occasionally one snags on something, and I have to struggle to clear it from the hook. When this happens, Javier steps in and removes it for me before it can be carried out of reach. Most of the livers are enclosed in a filmy membrane, intact and smooth yet soft and yielding to pressure. Every so often a liver has a deep gash or the entire outer membrane is ripped away. These livers are warm to the touch, even through gloved hands, and they give way and crumble at the slightest pressure. Sour odors rise from these livers, hitting my naked face.

Along the thick edge of each liver is a tough white strip—the coronary ligament—which provides a target for the stationary metal hooks. Unlike the rest of the liver, which simply rips if impaled, this strip sustains the weight of the liver on the hook. The strip is slippery and folds in lengthy layers around an equally slippery hole, into which it is possible to insert almost the entire length of my index finger. I aim for this hole when I push the liver onto the cart hook.

After a few weeks of this work and tens of thousands of livers, I will be able to tell almost immediately by the subtle vibrations of each liver whether it has been successfully impaled. If the liver slides on too readily or there is a ripping sensation or slicing, I will know without looking that the hook is cutting through the flesh and the liver will fall back into my hands or onto the floor. If there is a satisfying solidness to the penetration, a kind of firm, singular rip that slows, stops, and does not give, the impaling has been successful, and I can move on to the next liver. Then there are the iffy impales—the times when the liver feels as though it both holds and rips at the same time. Even after taking my hands away, I must continue to stare at these livers for another four or five seconds to see whether they will slip off the hook or stay on. Most stay, but at the slightest downward feint, my hands will lunge back around the liver, ready to withdraw it from the hook and try for a better position.

As days on the liver line stretch into weeks and months, my body settles into a repertoire of tactile sensations. In addition to the feel of hot livers and cold wet rags, there is the strangeness, which becomes familiarity, of having my face and neck be the only exposed parts of my body. One day, after something half-liquid and half-flesh spurts from a liver into my left eye, I request a pair of clear plastic safety goggles from Oscar, who after several amicable lunchroom conversations about his winnings at a casino, has decided that I am a "good man." This leaves only my forehead, cheeks, nose, mouth, chin, and the top of my throat exposed to the outside air. Every other part of my body is covered. I wear long underwear. Over these, a pair of jeans and a pair of water-resistant track pants. And after several weeks of leaving work soaked to the skin as a result of water splashing off the bloodied carts when I use the cooler hose to rinse them down, I request and receive bright yellow rain gear like that worn by the two men at the cooler entrance. Over my feet, two pairs of socks, cotton and wool, then the company-supplied green rubber boots with steel toes and black soles. On the upper trunk, a short-sleeved T-shirt, three long-sleeved T-shirts, a sleeveless hooded shirt, a sweater with hood, the bib of the yellow rain gear, and, over all this and hanging down to my knees, the company-supplied white frock with silver snap buttons. On my head, over the hairnet and under the white hard hat, the gray hood of my sweater is pulled forward as far as possible to cover my ears. Though restrictive, these layers soon feel like a second skin, and I take comfort in the warmth they provide as long as I keep my body moving throughout the day.

The feel of the cold steel cart, the vibrations that travel along it as the aging wheels creak, drag, and then finally spin against the concrete floor, also becomes familiar, as does the feel of sharp steel bars pressing into my palms as my burning legs stretch back and my elbows lock to give the initial push to move it with its load of a hundred or more livers, each weighing between ten and twenty pounds, forward and out of the way. I become used to the feel of the cooler hose, a live, writhing snake when the water is turned on, and the splash of blood and water off the cart as it is sprayed clean after the livers have been unloaded.

Also familiar is the unwelcome feel of water trickling down inside the cuff of my rubber glove and soaking the white cotton glove underneath it, and the sinking feeling in my stomach that accompanies this as I realize that once again I have dipped the rag down too deeply into the bucket and will now have to work with wet hands for the remainder of the day. I become used to the snag and rip of a rubber glove catching on the sharp edge of one of the hooks and the feel of hot blood and cold water finding this tear and rushing through to the cotton glove and bare hand underneath.

More painful is the feel of a half-carcass as it hits my arm on its path down the decline and through the cooler door. Moving rapidly and weighing over four hundred pounds, these swinging projectiles can leave a bruise after even momentary contact. Full contact can send a person sprawling backward onto the floor. Everyone tries hard not to stand too close to the carcasses' pathway, but no one can completely avoid the forgetfulness and misjudgment that come with fatigue, monotony, and the stress of completing tasks quickly in a confined space.

A constant source of pain is the feel of concrete through steel-toed rubber boots. The floor seems to get harder and harder each day, and by Thursday of each week it has solidified into an unyielding adversary intent on hurting your knees and breaking your spine.

Accompanying all these is the feel of other humans, palpable even through all the layers of cloth, plastic, and rubber: the emblematic clenched fist meeting the knuckles of another clenched fist, eyes locked through the clear plastic of safety glasses in the cooler workers' universal silent greeting, an acknowledgment of another's shared presence; the feel of the plastic raincoated arm of the taller mustached man at the entrance to the cooler door, sometimes around my shoulders, sometimes around my waist, whenever I stray over to talk to him. This is a common sight in the cooler: two men, each with an arm around the other in a small gesture of humanity and warmth, standing side by side with cheeks almost touching as they struggle to hear and be heard.

And just as the body settles into a familiar tactile rhythm, so too does the ear acclimate to its new aural home. After my first ten hours in the cooler, the rumbling of the fans continued in my head long into the night. The day is an interminable aural barrage, deep and throbbing. The cooling system is the constant background noise for everything that happens. The long, sharp *cling, cling, cling* of the half-carcasses straining with the force of gravity on their pulley wheels punctuates the dull background roar of the fans.

Down the overhead rail the half-carcasses slide, pulled along by gravity before slamming into a hydraulically operated, timed metal pin that juts out at the bottom of the stairway just before each half-carcass reaches it. When the pulley wheel hits the metal pin there is a loud chink, and the carcass whips forward then backward in a sharp movement that starts with the raised hindquarter, then works its way down like a giant spasm to the headless neck. The metal pin retracts, and the carcass continues on its gravity-powered run through the doorway of the cooler, more slowly now but still powerful enough to knock down an unwary worker. *Cling, cling, cling: chink. Cling, cling, cling: chink.* This is the rhythmic beat of the half-carcasses descending the stairs.

Accompanying it is the *hiss, hiss, pop* of the two yellow-jacketed mustached men as they use a small blue hose to blow compressed air at the condensation on the overhead rails and sprockets. The air hisses as it leaves the hose and makes a diffuse *pop* as it hits the metal rails. *Hiss, hiss, pop. Hiss, hiss, pop.* Finally, there is the *splink, splink, splink* of water hitting the bloodied metal carts as they are rinsed.

The subdued roar of the cooling fans, the *cling, cling, cling: chink* of the carcasses, the *hiss, hiss, pop* of the compressed-air hose, and the *splink, splink, splink* of water hitting metal join in a thousand different combinations and variations to form the mechanical soundscape of the cooler.

Human whistling pierces this mechanical cacophony at sporadic intervals. The urgent whistles start high and dip low before climbing high again. Is someone in trouble, in need of help? Have you made a dangerous mistake? Is there a supervisor or inspector coming your way? The less urgent whistles, short and sharp, start low and rise briefly. These are the whistles of someone who needs to move past making his presence known. The teasing whistles rise and fall like a catcall in response to someone slipping and falling or losing his grip on a liver and letting it fall to the floor. And occasionally one hears the aimless, meandering whistles of someone who is repeating a fragment of a tune to carry him through the next hour.

After Javier and I fill a couple of carts, a tall, broad-shouldered man in his twenties in a white frock and hard hat joins us at the liver line. He has a wide pale face, and when he talks to Javier his mouth shows several missing teeth. Moments later Javier hands him the rag and gives me a short wave before leaving the cooler. After I take another liver off the moving hooks, the new man—I learn later that his name is Carlos—hands me the rag and signals that I should wipe the hooks.

Where Javier and I have been carefully using two hands to guide the livers off the moving hooks and onto the cart hooks, Carlos often lifts the livers off the moving hooks with a single hand, hooking his index and middle finger into the tough, slippery opening in the fatty strip on the thick side of the liver, pivoting on his feet, and flinging the liver onto the stationary cart with surprising accuracy and force. And unlike Javier, Carlos does not wait for the livers to begin their ascent back up the cable before unimpaling them. Rather, he stands closer to the cooler door and retrieves the livers while they are still descending. From what I can see, this has the advantage of leaving much more time for the person with the wet rag to clean the bloodied hooks, but it also means stepping under the moving cable of ascending hooks. Carlos is fastidious about the cleanliness of the hooks. Several times, he grabs the wet rag out of my hands and wipes the hooks that I have already cleaned, shaking his head back and forth as he does so. I quickly learn not to leave any blood or flesh on the hooks.

Carlos treats the liver carts with more precision and deliberation than Javier as well. When there is a small break in the line of descending livers, Carlos retrieves several more empty carts, rather than one at a time, and lines them up near the cooler wall within easy reach. He also repositions the cart he is filling up so that its length is parallel to the overhead cable, permitting him to span the gap between the moving cable and the stationary cart simply by turning at the hips. After the hooks on the bottom two crossbars of one side of the cart are filled with livers, Carlos turns the cart 180 degrees and loads the hooks on the bottom two rows of the other side, then continues loading the hooks on the top two rows of this side until it is completely filled with livers. He then rotates the cart again and loads the top two rows of the first side. After the cart is completely loaded, Carlos pulls it about fifteen feet away, where he arranges it in a neat line, end to end with the carts Javier and I have already filled. We continue like this, Carlos loading the livers and I wiping the hooks, until the overhead cable stops suddenly and Carlos turns to me with one fist next to the other, then pulls them apart. "Break!" he yells.

I follow Carlos out of the cooler and through a side door at the bottom of the stairway into a narrow room about five feet by ten with exposed metal and copper pipes snaking across the cinder-block walls. The room is full of men taking off yellow rain gear and gloves, and I follow Carlos's lead in taking off my own frock and gloves. Then we are trotting, almost running, through some sort of machine shop. Large gears, pipes, welding equipment, and masks lie on the floor of the room, which smells pleasantly of grease and metal. At the far end we pass through another set of double doors into a dark, dank passageway with exposed pipes along the ceiling. There are several rooms off this passageway that reek tremendously, both of concentrated chemical and of organic odors. At the end of the hall, we barge through a third set of double doors into a hot, brightly lit room vibrating with a painful, high-pitched whine, its walls stacked almost to the ceiling with piles of flat cardboard. A white conveyor belt with vertical ridges runs up from the middle of the floor through the ceiling of the room, and a large black man in a faded T-shirt stained dark with sweat sits on a stool near the conveyor folding boxes from the flat cardboard and placing them between the ridges. His hands move mechanically, as if detached from the rest of his body, and he stares blankly into the middle distance even as we rush past him. At the end of the room is an area with a computer on a white desk surrounded by metal shelving and enclosed by wire-mesh caging.

A clock against the wall in the cage reads five minutes past nine, and a thin, oldish white man in a dark-blue uniform sits on a stool under the clock filling out a crossword puzzle. Like the black man, he does not look up as we pass. Next to the cage is a metal door with a red-lettered sign reading, "Emergency Exit." Carlos strides to the door and slams his open palm on the silver metal crossbar, and the warm, blinding June sunshine is upon us.

White-helmeted and frocked men and women are everywhere. Some sit on backless black plastic benches pushed up against the slaughterhouse wall, others squat close to the ground, their backs against the wall for support, and still others are scattered around the parking lot in small groups of two to five. Their boots are black or brown and made of leather, and fleshy white specks of fat cling to them and to the cuffs of the workers' jeans. Like me, they wear hoods under their hard hats. As Carlos and I join them the door to the main employee entrance pushes open and another crowd of workers, mostly men, surges through. Clad in the green boots that I wear, these workers do not have white frocks, and their various colored short-sleeved T-shirts are splattered with fresh blood. Some wear white hard hats and others gray.

Three vendors sell food out of trucks and vans surrounded by large groups of people picking out paper- and plastic-wrapped food and pushing money into the vendors' hands. Carlos heads for the middle vendor and I follow. In the back of a covered truck, set in a wire display case with glass doors, are Styrofoam plates with scrambled eggs, bacon, muffins, bread, and croissants, as well as meat- and potato-filled pastries wrapped in wax paper. I pay $2.50 for a pastry and soda and follow Carlos to the curb, where he is knocking a clenched fist against the outstretched fists of three other men in white frocks and hard hats. He sits beside them on the curb, and I have barely finished unwrapping my pastry when he is already rising to throw his plate into a trash bin. I stuff the pastry into my mouth and manage to swallow half my drink before I have to throw everything into the bin and trot to catch up to Carlos as he disappears through the emergency exit door.

The clock in the wire cage room reads twelve minutes past nine. I follow Carlos through the maze of rooms and hallways back to where we have hung our frocks and gloves. There are already about four or five men in the room putting their gear back on. They file out of the room while I still struggle with my gloves, earplugs, and frock—I have put on my frock and two layers of gloves before I realize that I will not be able to insert the earplugs with my gloves on, and so I have to take these off again, insert the earplugs, and then put the gloves back on. By the time I open the door and step into the area at the bottom of the stairway, half-carcasses are already clinking down the stairs and whizzing through the cooler door. Directly overhead, livers are already turning the corner at the higher sprocket and disappearing behind the cooler wall. Carlos raises his index finger and shakes it at me, motions to the moving liver line, and glances at an imaginary watch on his wrist. As we settle back into our prebreak rhythm, the warmth of the outside sunshine gradually yields to the unstoppable cold and the gobbled pastry and half-finished drink turn to poison in my stomach.

Four hundred livers later, Javier reappears in the fog of the cooler's entrance accompanied by a short stocky older man with sparse grayish hair on his upper lip and eyes that glance quickly around the room, never resting on anything for more than a second. Like Carlos and me, he wears a white hard hat and frock. Javier motions for the man to join us and, leaning close to Carlos's ear, speaks a few inaudible words. Javier turns to leave, and Carlos hands the man a second rag, directing him to stand next to me and wipe the hooks. I nod a friendly smile at the man, who nods curtly back at me, then turns his attention to the hooks. The man and I stand shoulder to shoulder, confined to the space between where the hooks turn the corner around the lower sprocket and where Carlos stands reaching up under the closer cable to retrieve the livers on the descending cable. The bucket of water sits behind us at our heels. With two of us wiping, there is little to do but wipe every hook, dirty or not, and the man scrubs each with such fastidious vigor that I soon feel obligated to do the same. "What is your name?" I say loudly into his ear, which I notice is uncovered by a hood or hat. He ignores me so I try again in Spanish. He responds by shaking his head, his full attention fixed on the hook wrapped within the wet rag in his hand.

After several more carts of livers have been filled, the overhead line stops abruptly, and Carlos makes the hand motion for "break"; we get a half hour for lunch. Taking off my frock in the warming room, I am finally able to talk to the new liver hanger. I learn that his name is Ramón, that he is fifty-nine years old, and that he is originally from Michoacán, Mexico. This is his first day working in the slaughterhouse. Before moving to Nebraska to join his two sons, both of whom work on the fabrication side of the slaughterhouse, he was a construction worker in California.

Soon I also get to know the two mustached men working near the entrance to the cooler. The taller is Christian. He is also from Mexico and has been working at the plant for four years, but he plans to return home soon; he has two daughters whom he has not seen since he left to find work in the United States. The shorter man is Umberto, and he has been working in the slaughterhouse for only a year. Previously he worked at a plant nursery in Omaha. Both Christian and Umberto are in their thirties, and both exude a certain air of gravitas as they stand at the entrance to the cooler wiping down condensation with their long-poled squeegees and tagging half-sides as they whiz into the cooler from the decline. Belying this gravitas is their fondness for playing practical jokes on each other and, after a week or two, on Ramón and me as well. These include sneaking up behind us and prodding us with their squeegees, throwing wads of fat at us, and tying rubber gloves to their poles and dangling them in front of us while we hang livers or clean the hooks.

A little farther inside the cooler is another pair of yellow-jacketed workers known as railers. The work of the railers inside the cooler is extremely demanding. Using clawed hooks anchored in an orange plastic base and held in one hand, the railers push, pull, and shove the four-to-six-hundred-pound half-carcasses into place. The physicality of the work shows in the age and physique of the two railers: both are young, in their late teens or early twenties. One, a stocky muscular white man named Tyler, is friendly and enthusiastic and immediately takes me under his wing, showing me, for example, how to roll up my earplugs before inserting them to make sure that they stay put. After time in prison on drug charges, Tyler now lives in a community house in Omaha and works in the slaughterhouse as part of a work-release program. Before his conviction, Tyler worked in a different slaughterhouse, also in the cooler as a carcass railer.

The other interior railer, Andrés, comes off as abrasive and aggressive. As soon as he sees me in the warming room, he calls me "Chino," the Spanish term for "Chinese," and this will be his name for me even after several weeks of working together. Tyler and Andrés relate to each other primarily through bravado and mimicked aggression, shoving each other into walls, punching each other, and challenging each other to push-up contests in the warming room. Although they refer to each other by a running litany of ethnic slurs (Andrés to Tyler: "white bread"; Tyler to Andrés: "wetback"; Andrés to Tyler: "white fatty"; Tyler to Andrés: "bean boy"), their mutual fondness is apparent. Their constant mock-fighting constitutes a source of entertainment and diversion for Christian, Umberto, Ramón, and me. It is not long before the six of us begin to form a sense of community, nurtured in part by the interactions made possible by the absence of constant red- and yellow-hat supervision in the frigid temperatures of the cooler.

On my second day, Carlos leaves Ramón and me at the liver line in order to lead a three-person team responsible for taking the chilled livers off the carts, wrapping them in plastic, and packing them, two at a time, into sturdy cardboard boxes. After the livers are boxed, they are stacked on pallets before being taken by forklift to the freezer room to be stored awaiting shipment. Carlos and his two liver-packing partners, Ray and Manuel, remain aloof from Andrés, Ramón, and the rest of us. In part this is because, unlike ours, their work schedule is not tied to the overhead chain. Ramón and I must work whenever the overhead line of livers is moving. Likewise, Christian, Umberto, Tyler, and Andrés must work whenever carcasses are coming down the decline on the overhead rail. Carlos, Ray, and Manuel, on the other hand, are constrained only by the need to make sure that enough of the hook carts are available for Ramón and me to hang livers on. Often they will work furiously for a period and then disappear for up to an hour on an extended break.

The relative freedom of the liver-packing team generates a certain amount of frustration for those of us chained to the line. Ramón and I cannot leave the overhead chain for even a minute lest the livers circle around the cooler and head up the decline and back onto the kill floor, an offense that can get us fired. Carlos, Ray, and Manuel often saunter by us on their way to or from one of their extended breaks, hands in their frocks as they whistle and watch us removing the livers from the overhead chain. Sometimes this general antagonism erupts into outright hostility over which group—the liver hangers or the liver packers—should be responsible for rinsing down the bloodied hook carts after the livers have been removed and packed.

No one wants to clean the liver carts. The work involves turning on a hose attached to a faucet protruding from the cooler wall and spraying the blood—which has congealed in the cold air of the cooler—off each row of hooks. Since the water cannot be sprayed with much force lest it splash onto the unpacked livers still chilling on the other carts, the person cleaning the carts also has to use a soapy rag to scrub the carts as they are being sprayed. The work leads to wet clothing and torn rubber gloves, which means cold, wet hands for the remainder of the day.

Carlos, the only one among the liver packers and hangers with any previous experience in the cooler, insists that the cleaning of the emptied liver carts has been, is, and always will be the work of the liver hangers. Ramón and I counter by arguing that when one of us is pulled away from the line to wash the bloodied carts, this leaves only one person to remove the livers and clean the hooks, a situation that leads to a frenzied back and forth between livers and empty hooks. Because it is Ramón and I, and not Carlos and his team, who need the clean empty carts, however, all the liver packers have to do to win the argument is nothing. By ignoring the bloodied carts after they have removed the livers, they force Ramón and me to rinse them when we begin to run out of clean carts.

The balance of power favoring the liver packers over the liver hangers shifts abruptly one day, however, when a USDA inspector issues the slaughterhouse a Noncompliance Report (referred to by slaughterhouse workers as an NR) when, because I am off washing a cart, Ramón touches a liver immediately after using a rag to clean one of the hooks. The inspector's reasoning for the citation is that the liver has been contaminated because the same glove that had been holding the contaminated cleaning rag touched the liver without first being washed.

Before this NR, the USDA inspectors have played a minor role in the day-to-day work of liver hanging. Christian and Umberto, standing guard at the entrance to the cooler, issue a high-pitched whistle whenever an inspector comes down the decline or through the warming room into the cooler, something that happens only a handful of times each day. Ramón and I take extra care not to drop a liver on the floor while the inspector is watching us work. Aside from this, the USDA inspectors do not have much impact on our work.

After the NR, the work of liver hanging suddenly attracts the attention of red-hat supervisors, green-hat quality-control workers, and the kill floor manager himself. Javier and James, our immediate yellow- and red-hat supervisors, caution us to try to avoid getting any more NRs. When we reply that we have simply been doing what we had been trained by Javier and Carlos to do, they respond that the USDA inspector is being "an asshole," but from now on we will have to stop wiping hooks whenever there is only one of us on the liver line. This potential solution to the problem strikes Jill, the green-hat quality-control worker, as insufficient, since the USDA inspector can still issue NRs if the liver hooks circulating back to the kill floor are visibly dirty.

Sensing an opportunity to score against the liver packers, Ramón and I mention to James that the primary reason we are left with only one person on the liver line is because the other person is constantly having to spend several minutes each hour cleaning the bloodied carts. The liver packers, we argue, have plenty of extra time left over after they finish a round of packing; would it not make more sense, in terms of avoiding future NRs from the USDA, for them to wash the carts after they empty them?

Much to our glee, James warms immediately to this argument, and Ramón and I trade high fives, give each other the thumbs up, and smile broadly as James walks over to speak with Carlos. It appears to be a highly contested conversation, with Carlos shaking his head furiously and James motioning to the carts and the liver line several times. The conversation ends with James pointing his finger severely at Carlos before turning and walking to the cooler door. "Okay guys," he tells us as he walks by, "those guys are gonna help you clean the carts from now on."

This turns out to be a hollow victory for Ramón and me because "help you clean the carts" translates in practice into Carlos ordering either Ray or Manuel to spray a few drops of water on each cart as it is emptied. Symbolically, though, Ramón and I feel the shift in power to be a major accomplishment, and it marks a key moment of solidarity in our working relationship. Christian, Umberto, Tyler, and Andrés—much amused by our recounting of the incident later that day in the warming room during the afternoon break—also celebrate it with us as a victory of those who are tied to the line over those who are not. What amazes me is how utterly absorbed, indeed obsessed, Ramón and I have become with this small and relatively insignificant struggle. Standing for ten hours a day in the damp cold of the cooler, we feel our small, symbolic victory in the battle of the liver-cart washing as an immense achievement.

And yet, if there are moments of victory, Ramón and I also experience many instances of near disaster. Early on, we learn the importance of rotating the liver cart each time one or two horizontal rows have been filled, rather than simply loading one side of the cart, then the other. On several occasions, after loading the liver cart to the top on one side, we try to turn it around, only to have the unbalanced cart crash on its side, sending fifty or more livers skidding across the wet floor. When this happens, Christian or Umberto issues a sharp, beckoning whistle, and Tyler or Andrés runs to the front of the cooler to help us rehang the livers while Christian and Umberto keep a lookout for supervisors or the USDA. If they signal that either one is on their way, Tyler and Andrés leave

Ramón and me to finish the job while they return to the middle of the cooler. The unspoken rule is that the six of us will aid one another but will not be required or expected to take the blame for anyone else.

A second hazard involves the deep, six-inch-wide drainage canals that stretch between two rows of hanging carcasses, including the front row near where we line the loaded carts of livers to be chilled. These canals are covered by perforated metal grates, and if one of the wheels of the liver cart is pushed onto the grate, its weight can cause the metal to buckle, tipping the cart on its side and sending a hundred or more livers skittering across the floor.

After several weeks of working together, Ramón and I develop a series of informal arrangements to make our work more bearable. Rather than take our breaks at the same time, leaving as soon as the liver line stops moving and rushing to return before it starts up again, we agree to stagger our breaks, with Ramón leaving five to ten minutes before the break is scheduled to start and returning sometime in the middle of the break to relieve me, while I return five to ten minutes after the scheduled break is over. This break staggering is crucial for managing, although not overcoming, the tyranny of being tied to the liver line. Learning that we live less than five minutes from each other, we also begin commuting to work together, taking turns driving. This often leads to visits to the bank, the barber, the grocery store, and the Department of Motor Vehicles together after work, and fosters a sense of solidarity both inside and outside the cooler.

But despite these developments, the possibility that my whole world is contained in this—that for ten hours each day, every day into an unforeseeable future, my horizons begin and end at the dirty-white cooler walls and the substance of my world consists of wiping off bloodied metal hooks with a cold damp rag and unimpaling and re-impaling warm livers by their posterior venae cavae—creates a black recess in my mind. Beneath the nervousness, anxiety, and small victories involved in learning the job, beneath the newness of the white frock and the green steel-toed boots and the double layers of gloves and ear protection, lies a terror of monotony. I have an unshakable sensation that the limits of my world are rapidly closing in.

There is a level at which everything about the experience of white-hat work in the cooler might be read as a struggle between submission to and rebellion against this imprisonment. External material factors are involved, of course. Small things—a carcass that falls off its pulley and thuds to the floor, a cart of livers that tips over, a supervisor's reprimand, a running series of pranks and jokes among the cooler workers, the enormous resentments generated over seemingly trivial and insignificant matters, intentional unintentionality and deliberate carelessness about the work, and even physical pain itself—all become events of gigantic proportion in the landscape of sterile monotony that threatens to engulf the person in the regularity and sameness of the cooler. The mind greets these minor events with a euphoria arising from a reminder of the inherent unevenness of life. Even as the lips emit the groans or complaints at the inopportuneness of a mechanical breakdown or curse at the incompetence of the mechanics, the mind delights in the rupture, slowly gathering strength for the interminable battle against monotony and the excruciatingly slow passage of time.

And when the unpredictable mechanical disruptions are long in coming and the flatness of the hours threatens to stretch into an unbearable eternity, then humans inject events of their own making. What the company rulebook refers to as "horseplay" becomes in this context essential to both psychological and physical survival. Pranks, jokes, random screams, whistles, and shouts, deliberate sabotage and surly insubordination, speeding one's work up frantically or slowing down to a pace that threatens to undermine everything, ongoing feuds over trifles invested with an emotional and intellectual energy beyond proportion: what are these if not monuments in a vast horizontal flatness? Here in the sunless, skyless humanmade confines of a near-freezing cavern deep within the bowels of the place where thousands of cattle are killed each day—here, perhaps especially here, there also exists an architecture and art of survival just as significant as those of the vast plains described by Wright Morris: "There's a simple reason for the grain elevators, as there is for everything, but the forces behind the reason, the reason for the reason, is the land and the sky. There is too much sky out here, for one thing, too much horizontal, too many lines without stops, so that the exclamation, the perpendicular had to come. Anyone who was born and raised on the plains knows that the high false front on the Feed Store, and the white water tower, are not a question of vanity. It's a problem of being. Of knowing you are there."[2]

At the rate of one cow, steer, or heifer slaughtered every twelve seconds per nine-hour working day, the reality that the work of the slaughterhouse centers around *killing* evaporates into a routinized, almost hallucinatory, blur. By the end of the day, by liver number 2,394 or foot number 9,576, it hardly matters *what* is being cut, shorn, sliced, shredded, hung, or washed: all that matters is that the day is once again, finally, coming to a close, offering a brief overnight respite from the roaring, vibrating totality that has come to encompass not only the knives, hooks, and machines that kill, rip, and tear apart the cattle but also the human arms, legs, and hands that operate these devices. This, too, becomes killing at a distance, laboring day after day hanging freshly gutted body parts from animals one never saw or heard or smelled or touched in life: the feel of freezing air through multiple layers of clothing; the smell of perforated livers rising to the nostrils in traceable steam; the humming, clanging, clinking, deafening mechanical soundscape; and the sight of liver after liver descending against a dull white wall, hour after hour, day after day, week after week until it constitutes an endless, infinite landscape in which the slaughtered cow has no place and against which every act of disruption, no matter how miniscule, becomes an expression of being, of knowing that you are still there.

VI
Killing at Close Range

Job Number 8, Presticker*: uses hand knife to make incision along length of the cow's neck, giving the sticker access to jugular veins and carotid arteries. Must take care not to be kicked in face, arms, chest, neck, or abdominal area by cows that are reflexively kicking, or kicking because they have not been knocked completely unconscious.*

"Guys, no more livers next week."

It is James, the red-hat supervisor in charge of the cooler, and he mumbles the words as he hands Ramón and me our Friday paychecks in the warming room while we take off our gear and get ready to head home. "But don't worry," he quickly adds, "we'll try to find some work for you guys. Just come back on Monday and we'll try to find something else."

We learn later that Russia or Korea—nobody really seems to know which—has temporarily stopped importing livers, and the management has decided to stop packing them until demand picks up again. And just like that, with two days' warning, Ramón and I find ourselves out of our jobs. Driving home, Ramón is anxious, asking me repeatedly what we are going to do, whether we are going to be fired, telling me he doesn't know how to use a knife and is not sure what other kind of work he can do. I commiserate, but internally I find the possibility of a break from the endless monotony of the cooler exciting and am hopeful that this will provide an opportunity to see a different part of the slaughterhouse. On the way home we stop at a Mexican grocery store, where Ramón picks up two tamales and a forty-ounce Miller Genuine Draft. I buy cheese, chips, and salsa. As we carry our bags to the car, Ramón asks me again whether we will have to look for another job. I tell him that I just don't know.

On Monday morning, Ramón and I stand nervously in the hallway opposite the kill floor office, hands in our pockets. Javier walks by whistling. We stop him, and he says he doesn't know what we'll be doing, but we should change into our work clothes and wait near the cafeteria. After about fifteen minutes of standing around in the hallway outside the cafeteria while kill floor workers rush past us to get to their stations, Ramón decides to check in the cafeteria to see whether someone is waiting for us in there.

Equally nervous at the thought of being out of work, and, like Ramón, knowing that each second that passes after the kill floor starts operating bodes ill for our chances of being given another job, I head onto the kill floor, where I see Bill Sloan, the son of the manager, talking with Ricardo, the red-hat supervisor.

"Do you guys know where I can work?" I ask.

"Do you have any knife experience?" Bill asks.

"No, but I can learn."

Ricardo shakes his head ominously.

"Guys, do you have anything outside in the chutes? I used to work on a ranch, and I'm good with live cattle," I plead.

Ricardo and Bill glance at each other, and Bill nods his head slightly. They both talk into their radios, then Ricardo motions for me to follow him through the clean side of the kill floor, where a line of white hats is standing ready for the first carcass of the day to make its way down the chain, and duck under a half-open garage door onto the gray-hat, hide-on, dirty side. There the line has already started, and the cattle swinging from the chain appear increasingly lifelike as we move down the line against the flow of production. Finally we arrive at a raised platform behind a gated area. A man in a black T-shirt leans over a waist-high barrier, a cylindrical silver gun in his hands, and every six seconds or so there is a *pffft, pffft* as the killing bolt strikes the cow, then retracts back into the gun in the man's hands, after which the cow falls forward onto the green conveyor belt below.

We climb the steps to the platform, and as we edge our way past the shooter, I can see the glistening sweat on his neck, even though it is only half past seven in the morning. Passing through an aluminum swinging door, we are suddenly beyond the walls of the slaughterhouse in a half-open enclosure. The odor is sharp and immediate, an acidic mixture of manure, urine, and vomit that stings my eyes and throat. Cattle, hooves clapping against the floor, push their way nose to rump in a continuous stream of hide up a chute with concrete walls about four and a half feet high and a foot thick. Two men stand on either side of the chute using metal-tipped prongs connected by wires to a live electrical line, plastic paddles, and a leather whip to push, nudge, and shove the cattle one by one through a dark hole at the end of the chute. There a conveyor catches them under the belly and lifts them off their feet, propelling them forward through a metal box to the knocking box, where the man in the black T-shirt stands ready to shoot them.

Covering the whole area is a low tin roof, only three or four feet from the top of my hard hat, dully lit by long fluorescent bulbs encased in plastic covers that are speckled with bits of feces. On either side of the chutes, three foot–wide concrete walkways are bordered by chest-high walls with plastic sheeting that stretches from the top of the walls to the frame holding up the roof.

From the hole in the wall that leads into the slaughterhouse, the chute descends at a steady slope for about fifteen feet before splitting into two parallel chutes. Known collectively as the serpentine, these chutes lead down into a circular area about forty feet in diameter called the squeeze pen. The cattle's movement from the squeeze pen up the serpentine is controlled by a series of gates and trapdoors. Beyond the squeeze pen, the cramped chute leads into a huge room with a peaked ceiling fifty feet high that is open to the air near the rafters. The enormous floor space is divided into pens with metal gates; some are empty, while others hold groups of cattle. This area is followed by the scale room, where cattle are weighed, and a raised concrete ledge where transport trucks unload their cattle.

Ricardo leaves me at the top of the chute in the charge of a short stocky man with a thin mustache named Camilo. In addition to the two of us, three other men work the area between the squeeze pen and the top of the chute. Directly across from Camilo and me a short thin man named Gilberto whistles and prods the cattle through the opening in the slaughterhouse wall. The squeeze pen and lower serpentines are worked by Fernando, a tall nineteen-year-old who immediately asks me if I belong to a gang, and Raul, a quiet man in his thirties who wears a blue bandanna in place of a hard hat and listens to a Walkman.

Gilberto and Camilo explain that our job is to "keep the line tight": to keep the cattle moving up the chutes and into the knocking box. Most of the cattle are moved into the primary serpentine chute, but five or six are also kept in the secondary chute in case there is a lull in the first chute. The cattle are organized in lots by seller, and when Fernando or Raul calls "Lot!" one of the upper chute workers uses an orange hide marker to write "LOT" on the back of the last animal from that group.

The size of a lot is determined solely by the number of cattle sold to the slaughterhouse from a single source: it can be as small as one or as large as several hundred. To be able to track the overall quality and age distribution of cattle from a source, lots are kept together when they are killed, and when the knocker sees the orange "LOT" on the back of an animal, he blows a loud air horn that signals to the supervisors and the workers responsible for keeping track of the lots that one lot is ending and another is beginning.

Camilo hands me an electric shocker and emphasizes that I should not use it when a USDA inspector is present. Because there are only two approaches to the chute area, one from the back through the cattle pens and the other from the front through the kill floor, the chute workers have developed a signal: a short whistle followed by a finger pointing at the eyes means that an inspector is coming over.

After a few hours in the chutes, it becomes clear to me

that both Gilberto and Camilo use the electric cattle prods extensively, sometimes sticking them under the animals' tails and into their anuses. The cattle jump and kick when shocked in this way, and many also bellow sharply. Gilberto uses the prod in almost rote fashion, shocking practically every animal, especially as they near the hole in the slaughterhouse wall that leads into the knocking box. Even when the cattle are tightly packed, with the nose of one animal pushed up against the rear of the animal in front of it—sometimes even with its head squished between the hind legs of the animal in front of it—Gilberto still delivers the electrical shock, often causing the cow to mount the animal in front of it.

Already caked in feces from their time in the feedlot, the transport truck, and the slaughterhouse holding pens, the cattle are packed so closely together as they push their way up the chutes that the defecation of one animal often smears the head of the animal immediately behind it. The impact of hooves against concrete splatters feces and vomit up over the chute walls, covering our arms and shirts, and sometimes hitting us in the face.

Running up the serpentine with swinging heads, the cattle are no more than a few inches away from us, separated only by the torso-high sides of the chute. Some poke their noses up over the chute wall to sniff at our arms and stomachs. I can run a bare hand over their smooth, wet noses, a millisecond of charged, unmediated physical contact. At close range, even caked in feces and vomit, the creatures are magnificent, awe-inspiring. Some are muscular and powerful, their horns sharp and strong. Others are soft and velvety, their coats sleek and sensuous. Thick eyelashes are raised to reveal bulging eyeballs with whites visible beneath darkly colored irises. I see my distorted reflection outlined in the convex mirror of their glossy eyes: a man wearing a hard hat, wielding a bright orange paddle. I look crazed, a carnival-mirror grotesque, upholder of a system that authorizes physical, linguistic, and social concealment to allow those who consume the products of this violence to remain blind to it. And what of the cattle, what of each of the twenty-five hundred creatures that are run through this chute each day? What do they see as they race by? What do they experience in the final moments before their deaths?

After months of the sterile, interminable monotony of hanging livers in the cooler, I am shocked by this confrontation with live cattle. Almost immediately, I resent Gilberto and Camilo for using electric prods (hotshots) on the animals, and after Camilo leaves the upper-chute area to take Raul's place in the lower chutes so that he can go to a doctor's appointment, I lean his prod against the back wall and pick up one of the orange plastic paddles instead. The rest of the day turns into an emotionally and physically draining blur of the "Hey, hey, yah, yah" call of the chute men, the slapping of the plastic paddle against hides, the bellowing and rearing of rolling-eyed cattle, and the incessant *pffft, pffft* of the knocking gun as it punctures one skull after another for hours on end.

Now that I am working on the dirty side as a gray hat, I am supposed to use the dirty men's bathroom and dirty men's lunchroom. We are the dirty men, no longer meant to interact with the clean men, the white hats. The chutes and pens, though, provide an informal gathering place of their own. Maintenance workers, supervisors, and USDA inspectors all use the semi-open area of the chutes as a place to take a cigarette break, to escape the confinement of the kill floor, to stand and talk while the stream of cattle runs by.

That afternoon, finished before the rest of the kill floor because the work of the chute men is done when the last animal is run through the chutes (it will be another forty-five minutes before that same animal is hanging in the cooler as two perfectly split half-sides), I wait for Ramón in the hallway. When he emerges from the clean men's bathroom almost an hour later, his hair and shirt are damp with sweat, and his clothes and arms are covered with small white specks of intestine. Driving home, he tells me he started out the day on the dirty side, unshackling the chain from the hind leg of the animals after they were attached to the overhead rail with a sturdier hook. He could not keep up with the work, and after the morning break they moved him to the gut room, where he threaded small intestines onto a coil that releases jets of water into them to clean them. Ramón complains that the gut room smells terrible, and that the work is hard, but he can probably get used to it. Then, after a few minutes of silence as he pulls bits of intestine out of his hair and tosses them through the open window, he says that he is going to look for another job since there is no future in this plant. As I drop him off, we agree that it will be better to drive to work separately from now on since work in the chutes begins at 6:30 and ends around 4:00, while work in the gut room begins after 7:00 and does not end until close to 5:00.

Next day in the chute the disagreement between me and the other chute workers over the use of the electric prods grows more heated. Camilo has replaced the knocker, and I am in Camilo's place in the upper chutes, standing across from Gilberto, using the plastic paddles to move the cattle. Both Gilberto and Fernando soon start yelling at me to use the electric prod. It is not just a matter of keeping the line tight, of making sure that there is little or no space between the animals, but also of keeping the line moving as quickly as possible so that the knocker and shackler can build up a surplus of stunned and shackled animals before the indexer spaces them evenly on the rails. Without the electric prods, the momentum of the line of animals is sufficient to move the cattle through the opening in the slaughterhouse wall into the knocking box, but not at the pace that the chute workers want. When shocked, the animals jump into the box, moving the line more quickly and reducing the probability of an animal's balking and holding up the line behind it.

Once, when the line moves too slowly for Fernando's liking, he sprints up the walkway from the squeeze pen, grabs the plastic paddle out of my hand, and shoves the electric prod into it. "You motherfucking pussy!" he yells. "Do your job and use the fucking hotshot!"

"Why?" I yell back. "What's the point of shocking them? They're all moving through the line anyway."

"The point is pain and torture," Fernando retorts, laughing. "Now do your motherfucking job and keep this line tight!" he screams, sauntering back down the walkway to the squeeze pen.

Across the chutes, Gilberto looks at me and shrugs before shoving his electric prod into the anus of one of the animals, causing it to kick back and then lunge forward into the animal in front of it.

"Why do you have to do that?" I yell at him.

He shrugs again, smiles, and keeps working. Furious, I repeat the question.

"Okay," he finally shouts back; "you wanna know why I use this?" He shoves the tip of the electric prod across the chute in my direction. "I use this because I like to have my work. And if we don't keep these cows moving through, they're gonna call us up to the office and we're going to get fired. That's why." Later that day we talk some more, and I learn that Gilberto has three children, aged twelve, nine, and six, and today is their first day of school.

By my third day in the chutes, after several warnings from Steve, the red-hat supervisor in charge of the area on the dirty side that includes the chutes and the pens, to "keep it tight," I too increasingly rely on the electric prod. The point of using the prod is not "pain and torture," in Fernando's mocking words, but rather avoiding conflict with co-workers and supervisors; in addition, once the abstract goal of keeping the line tight takes precedence over the individuality of the animals, it really does make sense to apply the electric shock regularly. Rather than electrocuting an individual animal, the prod keeps a steady stream of raw material entering the plant, satisfies co-workers and supervisors, and saves me from having to expend the energy it takes to move the animals with plastic paddles.

I try to take advantage of my proximity to the knocking box to learn something about the work of shooting the animals. One of the red-hat supervisors is temporarily manning the knocking gun for Camilo, and I ask him whether I could be trained to do that work. He says, "Yeah, I'll train you later. Now get back there and keep the line tight."

Later in the day, when Camilo is back at the knocking box, I ask him to teach me how to do the job. He tells me there are different controls in the knocking box area: one button powers the entire system; a lever controls the conveyor that runs under the animals after they enter the knocking box and lifts them off their feet; a second lever controls the side walls that move in and constrict the animal to keep it as still as possible before it is shot. Finally, there is a control for the overhead chains, which lift the cattle off the lower platform once they are shot and shackled.

The cylindrical gun is suspended in the air over the knocking box's conveyor, balanced with a counterweight and powered by compressed air supplied via a yellow tube. Camilo tells me that using it is not easy: the knocker has only one shot, and although the animals' bodies are restrained, their heads thrash wildly. It takes a combination of patience and good timing to hit an animal squarely in the skull about three inches above the eyes.

After shooting a couple of cattle, Camilo motions for me to take the gun. I do so while he controls the conveyor and the side restrainers. I am so focused on the gun that I do not even notice the animal that comes through on the conveyor. Its head swings back and forth wildly, eyes bulging. Then it stops moving for a moment, and I hold the gun against its skull and pull the trigger. Nothing happens. The gun has to be pressed harder against the animal's skull for the safety to be deactivated. I press again, harder, and pull the trigger. The gun recoils in my hands, and I see a hole in the animal's skull. Blood sputters, squirts, and then begins flowing steadily from the hole and the animal's eyes roll up into its shaking head. Its neck is extended and convulsing, and its tongue hangs out the side of its mouth. I look at Camilo, who motions for me to fire again. I shoot, and the animal's head falls heavily onto the conveyor below. Camilo advances the conveyor and the animal drops onto the lower conveyor, where it is shackled. There is already another animal in the knocking box, head swinging and eyes large in terror. I shoot two more animals, then Camilo takes the gun from my hands, warning, "They're looking at us." Two red-hat supervisors are standing farther down the kill floor, gesturing for me to return to the chutes.

Back in the chutes, Fernando asks, "Why you out there

doing that? You want to be the knocker?" When I say maybe, he responds, "No, you don't want to do that. I don't want to do that. Nobody wants to do that. You'll have bad dreams." This is the same man who told me that the point of using electric prods was "pain and torture."

Fernando's reaction turns out to be common. In the lunchroom, heating up my food, I talk to Jill, one of the two quality-control workers. We know each other from earlier conversations about dealing with the USDA inspectors when they watch the liver-hanging work.

"So, are you working in the pens?" she asks.

"Yeah."

"How do you like it? Do you like it more than the livers?"

I shrug noncommittally.

Jill holds her nose.

"Yeah, it smells pretty bad out there," I agree, then ask, "Do you know when the livers are going to start up again?"

"No, I don't know."

"I want them to train me to do the knocking," I offer.

She looks up, surprised. "You want to be a knocker?" Her voice is incredulous.

I shrug again.

"I already feel guilty enough as it is," she says.

"Do you really feel guilty?"

"Yeah. Especially when I go out there and see their cute little faces."

"Well, basically if you work here you're killing cattle," I say defensively. "I mean, aren't we all killing these cattle in one way or another?"

There is an uncomfortable silence.

"How long have you been working here?" I ask, shifting

the conversation, and I learn that she has been at the slaughterhouse for three years. She has taken classes to qualify for a USDA inspector's job, but does not want to apply for one because the work involves traveling and she has three small children at home.

The next morning, I am at work early for the free annual employee checkup provided by a company called Healthy and Well and paid for by the slaughterhouse. As an incentive to come to work an hour early, have your blood drawn to check for cholesterol levels, do a flexibility test (you sit with legs extended and see how far forward you can reach), have your blood pressure taken, and fill out a short questionnaire about your eating, sleeping, drinking, and smoking habits, the company provides a free breakfast of scrambled eggs, milk, juice, cereal, bananas, grapes, bagels, and cream cheese.

Rick, the safety coordinator, is responsible for enrolling employees for the checkup, and I sit across from him with a plate of scrambled eggs. When I tell him I want to be trained as a knocker he coughs on his eggs, then after a few minutes says, "You seem like the kind of person who would be really good for a desk job." It fits with a running conversation we have been having in which Rick has been encouraging me to start taking classes at the community college nearby and start looking for some other kind of work.

Later, I see Christian, Umberto, and Tyler, the railers from the cooler, and I join them. Christian and Umberto need to start working and eat and leave quickly. When I tell Tyler I shot three animals with the knocking gun the day before, he urges me to stop. "Man, that will mess you up. Knockers have to see a psychologist or a psychiatrist or whatever they're called every three months."

"Really? Why?"

"Because, man, that's killing," he says; "that shit will fuck you up for real."

I have no opportunity to become a knocker because my next day in the chutes, the fourth, is also my last. The day begins poorly. I am late to work, leaving me five minutes to get my gear on and get out to the chutes. Within minutes, an animal kicks up a big chunk of excrement that hits me squarely in the right eye. It stings, and I rinse it out with water from the sink at the knocker's stand, but I am worried about infection. Despite my increased use of the electric prod, Gilberto and, especially, Fernando continue to yell at me to "use the fucking hotshot," "watch your fucking side [of the chutes]," and "turn off the fucking fan." This last concerns an ongoing fight between Fernando and me over whether to turn on a large circular fan meant to provide some air circulation in the suffocating confines of the upper-chute area.

An hour into our work, a large brown heifer collapses in the knocking box just before it reaches the conveyor belt, blocking the passageway and shutting down the production line. Four USDA inspectors arrive, along with Roger Sloan, the kill floor manager, and his son Bill. They shoot the animal with the portable handheld knocking gun and attach a cable to its front legs. A winch drags the cow through the knocking box by its legs, clearing the way for the killing to resume.

Moments later, another animal collapses just inside the passageway leading from the squeeze pen into the primary chute. Steve, the red-hat supervisor, comes out, looks at the downed animal, and tells Raul and Fernando to route the remaining cattle up the secondary chute. A few minutes later Miguel, another red-hat supervisor, comes over and after talking into his radio orders everyone to take an early morning break. After the last animals are moved through the chute into the knocking box, the line is shut down, and the downed animal is shot with the portable knocking gun and dragged by the winch through the chute into the slaughterhouse.

Unbelievably, forty minutes after we return from morning break a third cow collapses in the chute. Lower down the chute, Gilberto has been shocking the cattle with the electric prod; I am using a plastic paddle to coax the animals through the hole into the knocking box. Shocked from the rear by Gilberto's electric prod, a cow mounts the steer in front of it. When the plastic paddle I am using to push the front-most animal into the knocking box spooks the animal behind it, the line of cattle in front of the cow that has mounted the steer pushes back, flipping the cow over onto its back and pinning it between the two sides of the chute. The cow struggles to right itself, but with the narrow passageway and downward slope slick with feces and vomit, it cannot get up. It soon lies still, breathing heavily and jerking its head back and forth, while the animals behind it come to a halt. Gilberto is furious, pointing at me with the electric prod and yelling, "You did this!"

Alerted by the stopped line of cattle, Fernando sprints up from the squeeze pen. "Good fucking job, Tim," he says when he sees the downed cow. Gilberto grabs a pair of metal rings off the wall behind him and tosses them to Fernando while I stand back against the wall. Fernando inserts the rings through the cow's nostrils, clamps them shut, and attaches them to a yellow rope, which he jerks heavily, trying to make the cow, now lying flat on its back, sit up and flip over onto its legs. Steve and several line workers from inside who have been alerted to the problem by the knocker join in the pulling. The pressure on the rope stretches the cow's nostrils until they are almost translucent. Finally, the men pull so hard that they rip the cow's nostrils and the nose rings fly out, hitting Juan in the hand. "Fuck!" he screams. The animal is thrashing back and forth in the chute now, its hide completely covered with the feces and vomit that layer the chute floor.

Richard, one of the maintenance workers who has been out in the chutes designing a compressed air–powered vibrator to use instead of the hotshots, is standing next to me against the wall. He looks appalled by what is happening. Steve motions to Gilberto to begin driving the cattle over the downed cow and raises two fingers to his eyes, signaling that all of us should be on the lookout for USDA inspectors. With electric prods Gilberto and Fernando push the remaining cattle over the downed cow, and they stomp on its neck and underbelly trying to escape the electric shock. Leaning against the wall, I look at Richard, who says shakily, "Man, this isn't right, running them other cattle over this cow like that. I'm not going to take part in this. I'm not going to stand and watch this." I nod my head in agreement, but both of us continue to stand against the wall.

After three cattle trample over the downed cow, I approach Steve and ask, "Do you really want to run the cattle over this one? If the other cattle break something in this cow, then it will never be able to get up." Steve ignores me but a moment later motions for Gilberto to stop driving the cattle over the downed cow, turns to Richard, and screams, his face only a foot from Richard's, "I'm going to get a cable and pull this beef through, and I want you to keep your fucking mouth shut about it. I don't want you to say nothing about it like the last time. Do you understand?"

Looking a bit stunned, Richard replies, "Um, yeah, I guess so."

"You keep your mouth shut," Steve says again, for emphasis.

Steve yells for some of the workers to get the cable and hook it up to the winch. But then his radio crackles with an alert from one of the red-hat supervisors inside the kill floor that the USDA is on its way. Yelling "Forget it!" he shouts at Fernando to get the water hose, which Gilberto hands to me and I hand to him, and begins hosing the hoofprints off the downed cow so that the inspectors will not be able to see that it has been trampled.

Within minutes, two USDA inspectors walk over. "What do you want me to do with this beef?" Steve asks them. The USDA veterinarian who usually inspects the live cattle says, "I want you to knock it and take it out the door."

"Which way, this way?" pointing down the chute back toward the pens.

"I don't care which way you pull it out, I just want you to take it outside. I don't want you to hang it"—meaning he does not want it processed.

Suddenly, Steve turns to me and orders, "You go see Ricardo in the lunchroom."

I assume that I will be fired because of the downed cow, as Fernando and Gilberto are in agreement that it is my fault. It is also possible that Steve does not want me present when the inspectors ask employees how the animal went down. I go look for Ricardo; not finding him in the cafeteria, I return to the chutes. The USDA inspectors and the downed cow are gone, and the cattle are moving through the chutes again. Roger and Bill Sloan are now in the chutes with Ricardo, and Gilberto is talking to them, his hands gesturing furiously. When they are finished talking with Gilberto, Bill Sloan turns to me: "What happened?" I tell him that the animal went down when it was pushed backward by the line of cattle in front of it, omitting any reference to the nose rings and the cattle being run over the downed animal. Roger, Bill, and Ricardo confer in a small huddle. As they leave, Roger turns to Gilberto and me and says clearly, "If this happens again, you two can both go home."

Roger and Bill return to the clean side of the kill floor, but Ricardo hangs back, talking with the knocker as he works. I approach him and ask whether the livers are going to start again on Monday. When he says he thinks so, I ask if I can be moved back to work on the liver line. "I really don't like being here," I say; "we have to use the electric prods too much." Ricardo tells me he will see what he can do.

Less than an hour later, a utility worker I have never met before enters the chute area and announces, "Someone is going home." Sullen and silent since our admonishment by Roger, Gilberto and I glare at each other. "I don't know who it is," the utility worker says, "but they told me to come back here because someone here is going home." Fernando, who has moved up the chutes with the last lot of cattle to be killed before the lunch break, points at me, laughs, and says, "Yeah, tell this motherfucker to go home."

After pushing the last few cattle through the knocking box, we go to the dirty side lunchroom, where Ricardo pulls me aside and tells me that Ramón and I are both being moved back to the cooler to get ready for the livers that will be starting again next week. "We have some extra guys, so since you don't want to be out there in the chutes anymore we're gonna switch you with the other guy so you can work with the liver guys again," he says.

I walk over to the clean-side lunchroom to tell Ramón, who is happy to learn that we will be working together on the liver line again. Just before the lunch break ends, James, the supervisor in charge of the cooler, tells Ramón and me to head down and spend the afternoon cleaning off the carts and hooks in preparation for the livers on Monday. "When you guys are done with that," he says, "just go to the box room and fold some boxes. But don't work too fast so you'll still have some work to do tomorrow [Friday]. You can leave at three, but don't let anyone see you, and I'll put you down for working the whole day. Then on Monday, everything is the same as before. You guys will start at seven hanging livers downstairs in the cooler."

In my four days working as a gray hat in the chutes, I drove no fewer than six thousand individual cattle into the knocking box, watched many of them get shot through the skull at close range, and shot three with my own hands. And although I spent most of this time in the chutes driving the cattle rather than as the knocker, Tyler's words—"Man, that's killing . . . that shit will fuck you up for real"—resonate deeply. "Fucked up" is exactly how I feel; it is how I would describe many of the chute workers, and it captures the rawness and violence of the perpetual confrontation between the living animals and the men driving them, myself included. What the experiences of Fernando, Raul, Gilberto, and Camilo suggest, though, is that three and a half days in the chutes, three and a half days in close proximity to the knocking box, is insufficient to understand what it means to do the work of killing at close range. Indeed, the experiences of the other chute workers indicate that there is some undetermined length of time, different for each individual, after which "fucked up" becomes routine, normal, and it is any sign of resistance to using the electric prods, to running live animals over a collapsed one, to piling the animals up like dominoes to be killed that becomes characterized as abnormal. Like Rick, Jill, Tyler, and many of the other kill floor workers, though, I do not want to traverse the terrain of routinization and normalization. The mythologizing of the work of the knocker—the almost supernaturally evil powers invested in the act of shooting the animals by the other kill floor workers, including, notably, the chute workers themselves—makes possible the construction of a killing "other" even on the kill floor of the industrialized slaughterhouse. It legitimizes and authorizes statements like the one made by Richard the maintenance worker, statements underscored rather than undercut by the fact that those making them are themselves contributing daily to the work of the kill floor: "I'm not going to take part in this. I'm not going to stand and watch this."

It is true. I would rather be cleaning hooks and hanging freshly eviscerated livers by the tens of thousands in the segregated confines of the cooler. The divisions of labor and space on the kill floor work to fragment sight, to fracture experience, and to neutralize the work of violence. But what I realize as I settle back into the hypnotic rhythms of wiping hooks and hanging livers by their posterior venae cavae is that this fragmentation, fracturing, and neutralization also create pockets of refuge, places of safety and sanity even here in the heart of the slaughterhouse.

The cooler and its monotonous rhythms are not only physically segregated from the correlates of killing by walls and partitions and the sterilizing effects of cold. More important, the cooler is also psychologically and morally segregated. Like Tom, Jill, and the other kill floor workers, I prefer to isolate and concentrate the work of killing in the person of the knocker, to participate in an implicit moral exchange in which the knocker alone performs the work of killing, while the work I do is morally unrelated to that killing. It is a fiction, but a convincing one, particularly for those already seeking to be convinced: of all workers in the plant, only the knocker delivers the blow that begins the irreversible process of transforming the live creatures into dead ones. Although the sticker technically kills the cow, it is unconscious by the time it reaches him. Only the knocker places the hot steel gun against the shaking, furry foreheads of creature after creature, sees his reflection in their rolling eyes, and pulls the trigger that will eventually rob them of life: only the knocker. If you listen carefully enough to the hundreds of workers performing the 120 other jobs on the kill floor, this might be the refrain you hear: "Only the knocker." It is simple moral math: the kill floor operates with 120 + 1 jobs. And as long as the 1 exists, as long as there is some plausible narrative that concentrates the heaviest weight of the dirtiest work on this 1, then the other 120 *kill floor workers* can say, and believe it, "I'm not going to take part in this. I'm not going to stand and watch this."

Months after I stopped working on the kill floor, I argued with a friend over who was more morally responsible for the killing of the animals: those who ate the meat or the 121 workers who did the killing. She maintained, passionately and with conviction, that the people who did the killing were more responsible because they were the ones performing the physical actions that took the animals' lives. Those who ate the meat, she claimed, were only indirectly responsible. I took the opposite position, holding that those who benefited at a distance, delegating this terrible work to others while disclaiming responsibility for it, bore more moral responsibility, particularly in contexts like the slaughterhouse, where those with the fewest opportunities in society performed the dirty work. My friend's position was the "120 + 1" argument, an argument replicated across myriad realms where morally dirty work is performed by a select few, out of the sight of the many who implicitly or explicitly authorize it but manage to evade responsibility for it by virtue of their citizenship, the taxes they pay, their race, their sex, or the actions of their ancestors.

But perhaps it is the preoccupation with moral responsibility itself that serves as a deflection. Perhaps there are at least some who would be willing to disavow the "120 + 1" argument and accept moral responsibility for the killing as a condition of benefiting from it, as long as they could continue to be shielded from any direct contact with or experience of it. In the words of the philosopher John Lachs, "The responsibility for an act can be passed on, but its experience cannot."[1] What might it mean, then, for all who benefit from dirty work not only to assume some share of *responsibility* for it but also to *experience* it: seeing, smelling, hearing, tasting, touching what it means to be the 1 in the 120 + 1?

VII
Control of Quality

Job Number 73, Tail Harvester: *uses hand knife to sever tail and hang it on offal line.*

"You look tired," Javier says to me on break one morning a week after I have been moved back from the chutes to hanging livers in the cooler with Ramón.

"I am tired. I only slept four hours last night."

"Why? Do you have another job?"

"No, I was reading. How about you?"

"Tired too."

"Do you have another job?" Javier shakes his head. "Well, why are you tired?"

"I don't know," his voice trails off. "Just tired."

"Hey, where's Julia?" I ask, referring to one of the women with a green hard hat who sometimes takes her breaks with us after the main group of workers is back on the line. "I haven't seen her in a couple days."

"I don't know," Javier replies. "She's not here. She hasn't been here for a while."

"I like her. She's nice and she's smart. Maybe she's on vacation?"

Javier shakes his head. "Yeah, something happened with her work and she left," he says. "She had a problem with one of the workers."

"What kind of problem?"

Javier looks at me seriously. "You know the QC [quality-control] job means you have to look," he says, pointing his index finger at his eyes for emphasis. "You have to look at what everyone is doing, and there was one worker who wasn't washing his hands, so she told Steve [one of the red-hat supervisors], and he told that worker to be more careful when she was around. So that worker, you know, got mad at her and went up to her and pushed her a little bit."

"Really?"

"Yeah, it was a big thing. She said she was going to call the police."

"Was the guy fired?"

"No, I think they gave him three days' suspension."

"So that's why she left?"

"No, later, on another day after that, Roger was upstairs in his office, and he saw Julia walk by one of the workers and push him."

"What? She just pushed him?"

"Yeah, and he wasn't doing anything to her. She just pushed him in the back. So Roger called her on the radio and told her to come upstairs."

"Did he fire her?"

"No, he told her, you know, you go home and you think about what you just did. And she never came back. So I think she quit."

"Wow."

"Yeah, so Roger asked me if I wanted to have the QC job, you know, to be the QC guy, but I don't wanna have that job."

"Why not?"

"It's easy money, you know," he explains, rubbing his thumb over his index and middle fingers, "but there's a lot of problems with that job because you have to deal with the inspectors, the workers, the supervisors, with Bill and Roger, and man, if anything happens you get in trouble with the company."

"Oh. Because it's the QC's job to make sure everything is okay?"

"Yeah," Javier says; "it's more money and I can do it, but I don't want to."

"Hey," I say jokingly as I put my yogurt lid in the empty container, "maybe it's time for me to get a promotion!"

Javier doesn't laugh with me. "Actually," he says, "me and James have been thinking about you for this job."

"Really?" After a minute I add, "Um, yeah, I could do that job. I can talk to people, you know, and I know some of the USDA guys."

Javier pulls his ponytail back up under his hairnet. "Yeah, you could do this job."

"Aren't there other people who want this job," I ask; "people who have been here a long time?"

"Yeah," Javier replies, "there are people who have been here a long time, but for this job you have to know how to speak English, you have to know how to read and how to write. A lot of people know more about this place than you do, but they can't do that."

I get up and put my plastic lunch bag back in the refrigerator. "So how do I get this job?"

"You have to talk to Roger or Bill."

There is no quashing the hope, however remote, that I might escape the monotony of the cooler. Numbness, the best method for condensing the hours of the day, has been usurped by optimism and anticipation. I tell Ramón that I need to get more soap but instead go looking for James, whom I find in the hallway. I tell him about the quality-control opening, and he says I need to talk to Bill, who has been out sick with food poisoning. While we are talking, Ricardo comes waddling down the hallway, and James tells him that I am interested in the quality-control job. Ricardo looks at me with raised eyebrows. I have trouble understanding what is behind the look—amusement? anger? fear? He says I need to wait until Bill comes back because Bill handles hiring. "What about Roger?" I ask.

"No," he says, shaking his head, "you have to ask Bill. If you ask Roger he will just tell you to talk to Bill."

I do not want to wait for Bill; I want movement, now. As I return to the cooler from lunch break, I catch the eye of Jill, the other green hat, a white woman in her twenties with long blond hair and green eyes. We have had conversations in the past, when the USDA inspector cited the slaughterhouse for improper cleaning of the liver hooks, and she waves for me to come over. I motion downstairs, signaling that she should come down to the cooler when she has a chance. Half an hour later, she walks into the cooler, and I take her into the warming room and tell her that I am interested in the quality-control position. She says that she already recommended me to Bill, long before Julia left, telling him that if they ever needed another QC, I would be a good choice. I ask who makes the decision, and she replies that both Roger and Bill do—Roger has more sway, but he often prefers to leave the smaller decisions to his son Bill. She asks me whether I know how to write "good," and says that Julia did not but not to tell anyone she said that. I ask whether she would be the one to train me, and she says that I would probably just follow her around for a week.

Back on the liver line, Ramón wants to know whether we are in trouble. I tell him there is a chance that I might become a green hat. "Very good," he says and turns back to the livers.

On our ten-minute afternoon break, I quickly peel off my yellow rain suit and white frock and take the back stairs to the upstairs hallway, where I see one of the red hats. I ask him where Roger is, and he points to a door marked "Kill Floor Office." I knock, and the door swings open. There are five or six men, all red-hat supervisors, lounging around a large conference table in the middle of the room. I notice James and Ricardo, but they do not meet my eyes.

Directly to my left, at a desk all his own, a large white man with a well-fed stomach is leaning back comfortably in an office chair, his hands clasped behind his head. He is wearing blue jeans, a pressed striped long-sleeved shirt, and gold-rimmed glasses. His thinning hair, free of the webbed markings impressed by a hairnet, is gray, almost silver. A spotless gleaming white helmet sits on his office desk.

"Mr. Sloan?" I ask.

"Yes," he answers neutrally.

"My name is Tim, and I work on the liver line in the cooler. I heard that a QC position recently opened up, and I want to apply for it." The words spill out of my mouth, tumbling over one another. Roger nods without saying anything. He holds my gaze. In the silence, the red hats look on.

"I have excellent reading and writing skills and I have good verbal communication," I continue. Before I came to this room, I did not know what I would say or how I would say it. I only knew that I wanted out of the cooler, out of the slow and endless descent of livers down the steamy decline.

"Well, you speak good," says Roger. I search for some sign of irony, but there is none.

"Thank you," I say. "I've been here over two months, and I already know most of the USDA inspectors. I think I would be good at this job. I have already talked to Jill about the job description and I know a little about HACCP [Hazard Analysis and Critical Control Point, pronounced "Hassip"] and things like that. I would really appreciate the opportunity to do this job." Roger looks more interested now.

"Well, you can't learn everything there is to learn in a day," he says. "I'll call Katherine [one of the vice presidents], and she'll send some paperwork over and we'll set up an interview."

"For tomorrow?"

"As soon as possible," he promises.

"Thank you," I gush. "When I first started working here two months ago someone told me that if I showed up and worked hard, opportunities would come up, and when this came up I thought I would seize the moment." I am genuflecting before this stranger with the gleaming white hard hat.

"Well, thank you for stopping in," Roger says, extending his hand. I shake it and leave, taking the steps down to the cooler two at a time. I am seven minutes late back to the line, and Ramón and I do not speak for the rest of the day.

The next morning, James walks into the cooler just before our morning break. "They want to see you in the kill floor office at nine-thirty."

"Who will take my place on the line?"

"I'll tell Javier to come down here after the morning break."

When Ramón returns from his morning break, I go to the locker room and take off layer after layer of my cooler clothing. I knock on the kill floor office door at 9:30, and James opens it, then returns to the big conference table, where he was eating a bowl of instant oatmeal. Roger walks into the room. "That stuff is supposed to be good for your heart," he says to James.

"Oh, yeah," James says. He quickly finishes his oatmeal and leaves the room. Roger looks at me and explains that we are waiting for a couple of people to come from the front office. Before long, two white women arrive. One appears to be in her early forties; she is tall and thin with a sharp nose and quick, inquisitive eyes. The other is shorter and perhaps five to eight years older, with a wrinkled face and glasses. Both wear crisp white frocks that look brand new and white hard hats with black decals reading "Visitor" across the front. Roger introduces the taller of the two as Katherine, vice president of technical operations for the company, and the other as Sally, the front-office quality-assurance manager. I shake each of their hands, and we pull up office chairs to sit in a circle.

Roger begins by recounting with great relish the story of how I had knocked on the office door the day before and introduced myself to him. "That just really blew me away," he says. "You never see kids that hungry anymore." Not exactly sure what he means by "hungry," I simply nod my head.

Katherine looks at me. "Tell us a little bit about yourself, your background, and your goals." It seems like a standard, open-ended interview question, and I give a quick rundown of the brief information I listed on my employment application. "I've never had a job like this before," I close by saying, "but I am a hard worker and a quick learner, and I would really appreciate the opportunity."

Roger and Katherine lean forward while I am speaking, smiling and nodding their heads from time to time. Sally sits back in her chair, passive and reserved, harder to read. When I finish talking, Roger repeats his earlier comment about kids not being "this hungry" anymore. He talks about the job as though it is already mine for the taking. "We like to promote from within rather than hire from the outside," he says. "You never know what you'll get when you hire from outside. But when you promote from within, you know you've got a good person because you've seen that they are loyal."

Katherine interjects, leaning forward in her chair and looking me in the eyes, "You'll find, Tim, that we try to operate like a family here. Loyalty is very important to us. If you are loyal to us and work hard for us, we will take care of you. If you have a problem with something or a question about anything, we want you to come and talk to *us* about it," emphasizing the "us" with a sweep of her hand. "We try to take care of each other here. We produce an extremely high quality of beef here, and part of your job as quality control is to help make sure that beef stays at that high level of quality."

"You won't learn everything there is to learn in one day," Roger says quickly, "and we don't expect you to either. Things are not ever 100 percent perfect here. Heck, I've been here for almost twenty years, and I still learn something new all the time. I don't pretend to know everything, and we don't expect you to know everything either. And Katherine's right that we're like family here. Heck, Katherine's been with us for almost twenty years now, James has been here eighteen years, and Michael, another one of the supervisors, is turning nineteen this year."

I ask about the job, what it is like.

"Quality control is a good job to have, and it was the last

person's loss that she decided to leave," Roger says. "It is hard work, but there are some perks," he continues. "You aren't tied to the line, you get a uniform, boots, a radio, a notepad, and you get to use the quality-control office. You get paid more overtime because you start at five in the morning and usually you don't finish until after five in the evening. You'll work with Jill, but you'll report directly to Bill and to me. If you have a problem with one of the workers, don't talk directly to them, talk to their supervisor. And," he adds, "you'll coordinate with Sally here, who handles things in the front office." It is the first reference in the entire conversation to Sally, and she nods her head, smiles tightly, and does not say anything.

I ask about money. "I'm in control of what people make here," Roger says. "I don't have to get anybody's permission to raise or lower someone's wages. For now, you'll stay at the same rate you've been at, but you could go up quickly if you learn fast, and I have a feeling you will."

The discussion turns back to my experience working on a cattle ranch, and somehow the topic of how the cattle are handled in the chutes comes up. "My philosophy," I tell them, "is that if you have to use brute force with a cow, then it probably means that the cow is smarter than you are." Katherine and Roger look quickly at each other, and there is nervous laughter from Roger.

"Oh yeah," Roger says quickly, "we send our kids to Kansas City for classes on animal handling all the time. There's this lady named Temple Grandin who runs those classes and she really gets your attention when she starts talking about cattle."

"Isn't she the autistic lady who designs slaughterhouses to help the cattle move through more easily?" I ask.

Katherine and Roger look at each other in surprise.

"Well, I love to read books," I say, "and there's this one

book about people who have different things wrong with their perceptions—like one guy can see only white and black and gray because of an accident—and one of the chapters is about Temple Grandin and the slaughterhouses she designs so that the cattle move up to be killed without getting stressed out. The design is called the stairway to heaven, I think. Anyway, it's a really good book, I think it's called *An Anthropologist on Mars* or something like that. I can get you the exact title and author if you're interested."

There's a silence in the room. I am on dangerous ground, but the conversation is intoxicating. It is more than just the desire to impress them so that I can get the job. After working in the complete silence of the cooler, I have an uncontrollable urge to show the management that we are not just stupid, mindless machines turning our gears day after day, to show them that we too have thoughts and feelings.

Roger breaks the silence with "If everyone could just sit and talk the way we are talking now, the world would be a better place. You know what the problem is with people these days, Tim? People just don't talk to each other anymore." Roger presses the call button on the black Motorola radio clipped to his belt. "James, you got a copy, James?" he says, winking at me as he does so. "James, can you come to the kill floor office please?"

"Well, James," says Roger, when he steps into the room, "I have good news and bad news. The good news is that you can keep Javier. The bad news is that I'm taking this young man away from you," motioning to me. "How soon can I have him?"

"Well, I guess he can start tomorrow," James responds.

Roger turns to me. "Be here at five tomorrow morning," he says, and stands to shake my hand. Katherine and Sally also stand up, and I shake their hands and thank them. I have the job. On the way back to the locker room to put on my warm clothing for the cooler, I find that my entire body is shaking.

Back in the locker room, I see one of the liver packers from the cooler, Ray, sitting on the bench taking off his boots. It is only about ten thirty in the morning. Ray is about twenty-three years old, slim, with a small beard. We have had short conversations in the warming room on afternoon breaks before.

"What's up?" I ask, and he tells me he is quitting. He has been paying someone, an American citizen, for the use of his social security number so that he can work in the plant. He has been paying $100 a week out of a paycheck of about $400, to a man named "Rick," who uses the money to pay child support. Now, Ray explains, Rick is demanding $150 each week or he will turn him over to the immigration authorities. "I can't pay that," Ray says, tears welling up. "I'm going to go apply at another plant now and use my brother's name."

"I'm sorry," I say, sitting next to him and putting my hand on his shoulder. "I'm sorry." But I feel more guilt than sorrow. Just two months into my job, I have been promoted to a semi-managerial position, a promotion based not on my experience on the job, my expert knowledge, or my seniority but simply on the fact that I know how to read, write, and speak English and am willing to pursue the position by telegraphing to the management that I am "one of them," that the world would be a better place if people (like me) could sit and talk to people (like them). Ray, who has worked in the cooler for longer than I have, has been giving up nearly a quarter of his wages just to be able to spend ten-hour days in freezing temperatures. And now, the price of admission unilaterally raised, he is forced to move on. We exchange phone numbers and promise to stay in touch, but when I call him a week later, his phone has been disconnected.

Back in the cooler, I tell Christian about Ray. He misunderstands me and thinks I am saying that the immigration authorities are in the slaughterhouse. "They're here?" he says. "*La migra* is here?" Fear spreads across his face.

"No, no, no," I try to reassure him, "nobody is here."

Later that day, on afternoon break, I tell the cooler workers about my promotion. There are smiles, backslaps, high fives, and expressions of "That's fucking great!" But not from Ramón, who says he feels sad. "It's good for you," he tells me on the car ride home, "but who's going to work with me now?"

The parking lot is deserted at 4:45 A.M., my new start time as a QC. The bulking mass of the slaughterhouse seems subdued in the predawn darkness, a slumbering monster. Even the smell seems less intense, more lingering odor than overwhelming force. At this hour there is no security guard checking identification cards in the entrance hallway. I walk to the locker room, where I put on my green rubber boots and white hard hat. I see Oscar, the supply-room manager, and Javier, the yellow hat, exchange brief greetings with them, then go to the quality-control office. Jill is inside reading a newspaper at the desk. She tells me that I will be trailing her all day; I can learn by watching what she does. She warns me to be very careful about what I say to the USDA inspectors, especially the head inspector, Donald, because he will try to trick me into saying the wrong thing so that he can issue a Noncompliance Report. NRs, she says, are reports that federal inspectors file against the plant, and they are "very, very bad" for the plant because if the plant gets too many of them, the USDA can impose fines or shut it down. In addition, the reports are accessible to the public.

There are thirteen USDA inspectors on the kill floor. Dr. Green is the inspector in charge (IIC); he is accompanied by another doctor of veterinary medicine, an antemortem inspector who examines live cattle prior to slaughter. A third inspector, Donald, has the title of consumer safety inspector (CSI) and roams the kill floor looking for food-safety violations. The CSI also inspects the food-safety documentation filled out by the QCs. Finally, there are ten line inspectors who rotate through stationary positions at the head inspection line (near circle 65 in figure 2 of chapter 3), the viscera or gut table (near circle 74), and the final trim rail (near circle 88). These inspectors do not have the power to issue NRs and must call either Dr. Green or Donald if they find a problem.

"Don't say nothing to Donald," Jill reiterates, "and if he asks you any questions that are work related, just tell him you don't know the answer; you're still in training." She hands me a radio and warns me to be careful to listen for when Roger calls and be sure to respond immediately if he does.

I try to ask Jill about the job, looking for a way to understand it conceptually before I have to start doing it. "You just have to watch what I do and learn," she tells me.

At exactly 5:00 A.M., Jill and I walk out onto the kill floor for what she calls a "pre-op," short for "pre-operational inspection." We will cover the clean side, and Javier will cover the dirty side. "Ever since Julia quit, Javier has been doing pre-op on the dirty side," Jill tells me. "He doesn't know how to do the clean side, so he just sticks to the dirty side. The dirty side is easy, but you're going to have to learn how to do both sides, so next week you can follow Javier on pre-op."

The kill floor is eerily quiet; equipment I have only previously seen covered in blood, chunks of muscle, and fat now gleams bright and metallic under the harsh white halogen lamps. The gut table, a five by forty foot moving table of joined metal plates on which the cattle are eviscerated before being split in two, looks like a futuristic runway. The overhead chain is still. Bereft of people and activity, the clean side of the kill floor can at last be seen as a single, yawning hangar, a whole. There is a silent wonder to the place.

Two men wearing sweatpants, T-shirts, and light-blue hard hats stenciled with the letters "DCS" approach us. One is in his late teens, thin, almost gaunt, with a face marked by acne. The other is older, perhaps mid-thirties, with long, curly hair worn in a ponytail. "Hi," Jill says, and they nod hello to her in return. These are some of the contract cleaners who work in two shifts beginning immediately after the kill floor shuts down for the day. Overnight, they remove the blood, fat, and bits of viscera, organs, and muscle from every surface and every piece of equipment on the kill floor. Their job is perhaps even more dangerous than that of the slaughterhouse workers; in addition to contending with industrial-strength chemicals that scorch and burn, these sanitation workers must also stick their arms and sometimes their entire bodies into dangerous machinery. Their hours correspond to the second and third maintenance shifts, when machinery is oiled and repaired. Sometimes the conflicting demands of repairing and cleaning the same machine can lead to horrific accidents. Like most of the slaughterhouse workers, the sanitation workers who do the actual cleaning make between seven and eight dollars an hour.

Jill switches her flashlight on and peers closely at hooks, rails, walkways, walls, and ceilings. She runs her bare hand under and over many of the surfaces, sometimes finding bits of fat under railings or grease from the overhead chain on conveyor belts. She points these out to the men, who contort their bodies heroically to reach into nooks and crannies. As we walk the plant floor, the two men engage in a sort of leapfrog, one cleaning the first area Jill points out, the other moving ahead with her to do the next.

As I will learn over time, Jill is playing a game of probabilities, making informed gambles based in part on her knowledge of where Donald is most and least likely to make a close inspection. She must weigh this knowledge against her experiences with individual sanitation workers: What are the spots they often miss? Where are they most dependable?

As we walk, Jill explains that we have forty-five minutes to cover our portion of the clean side of the kill floor. On any given morning, Donald can turn on the light above the USDA office door, which means that promptly at 5:45, he will emerge from the office, clipboard in hand, and conduct an inspection of randomly chosen areas or equipment in each of four pre-op zones on the kill floor. Zones 1 and 3 include the dirty side and the foot and gut rooms, where Javier is working. Zones 2 and 4 include the clean side, the decline, and the head table areas, where Jill and I are.

The pre-op walk is a constant shifting performance. In the first iteration, there are three teams: QCs, third-shift maintenance workers, and sanitation workers. The QCs play the role of audience while the maintenance and sanitation workers are the performers who know which areas in the plant have been unsatisfactorily cleaned or repaired and which were rushed because the crew was short on time or workers. It is in their interest to hide these areas from the QCs, insofar as possible, since discovery will mean more work for them and an admission that they did not do the job correctly the first time around.

As we walk the kill floor, Jill speaks frequently into her radio, calling either Lance, the white DCS sanitation supervisor, to identify a dirty piece of equipment or the third shift of plant maintenance, which is staffed by a white man named John and six of his brothers and cousins.

Lance is slim and wiry, with a ragged ponytail and several missing front teeth. He has a small windowless office near the dirty-side lunchroom off the kill floor, and by his own account spends most of the night chain-smoking in that room while his Latino employees clean the kill floor. Like many of the white people who work in the slaughterhouse industry, he does not do the actual work himself, but if there is a problem requiring a face and a communicator, he shows up to represent the work that has been done. When called on the radio he will inevitably respond, "I'll be right there," and will appear moments later holding a long extendable metal pole with a small scrub brush on the end. If the problem is small—a grease spot or a small area of dried blood—he will clean it himself. If it is larger, he will typically turn to one of his employees with an exasperated face and make pointing motions with his hands. They will usually respond with a shrug and an exasperated face back at him before cleaning the area.

John, the third-shift maintenance supervisor, is more often than not in a terrible mood. Any cleaning that requires a piece of equipment to be moved or disassembled is supposed to be maintenance's responsibility. The sanitation, or "DCS," workers, as they are called, are not permitted to remove or modify equipment. Almost every morning, there are overhead light covers with water trapped in them (an NR offense, since standing water can harbor harmful bacteria) or hoses that are leaking oil or hydraulic fluid onto conveyors (also an NR offense). When asked on the radio to open a light cover or check a hose for leaks, John's response is usually, "Do it your fucking self"; five or ten minutes later, long after the QC has moved on to a different spot, one of the maintenance employees on his shift will appear to do the work.

There is a running battle between the DCS and maintenance workers. Because the line demarcating what falls under the responsibility of sanitation versus maintenance is unclear, each often responds to a problem by telling the QC to call the other. Lance constantly complains that the equipment was cleaned but maintenance came in later and oiled or greased it. Maintenance, on the other hand, often accuses the DCS of claiming that there are oil or hydraulic leaks when the sanitation workers have not adequately cleaned the area. In forty-five minutes it is impossible even to inspect visually all the surfaces and equipment in the two zones of the kill floor, much less check them thoroughly. There are sixty different equipment areas in each zone, giving the quality-control worker an average of 22.5 seconds to look at each area, not including the time it takes to walk between locations. Some surfaces may appear to be clean but reveal a coating of fat when a QC runs a fingernail over them. Even worse, a large chunk of kidney or muscle or a pool of blood, grease, or hydraulic oil might be lurking in a corner, behind a piece of equipment, or stuck on the ceiling or wall. Some of the equipment, like the "185," an enclosed metal cabinet that sprays the 185-degree water on the carcasses as they pass through, has no internal lighting and contains dozens of tubular parts located at various heights ranging from ground level to more than fifteen feet. Dust, grease, or blood on the top of cabinets can also be cited by the USDA, yet inspecting these would require a ladder and take several minutes for each one.

Jill tells me that Roger and Bill used to give the QCs only

fifteen minutes to make their pre-operational inspections, but after a rash of NRs from the USDA for dirty equipment prior to the start of production, Roger agreed to increase the time to forty-five minutes. With this increase in time, however, his expectation is that there will be no NRs issued because of pre-operational deficiencies, increasing the pressure on the QCs to catch every possible offense before it is seen by the USDA inspector.

At 5:40, everyone listens carefully to the radio for John, who has the job of announcing whether the light is on above the USDA office. If it is on, the radio crackles with a dramatic, "We AAARE walking. We AAAAAARE walking." If it is off, he says, to everyone's relief, "We are NOOOOOOT walking. We are NOOOOOT walking."

"We AAAARE walking" comes over the radio at 5:45, and Donald walks out of the USDA office and heads to the maintenance shop with four clipboards in his hand, one for each of the inspection zones. Each clipboard contains a numbered list of locations and equipment from which he randomly selects three for inspection. Maintenance disconnects the power to the three pieces of equipment or locations and, where possible, locks them (a standard safety procedure known throughout the manufacturing industry as a lock out or tag out). The keys to these locks are placed in a red box, which Donald then locks with his own lock. The contrast between the safety precautions taken for the USDA inspector and those for the sanitation, quality-control, and maintenance workers is startling.

Donald walks toward the first piece of equipment on his list, followed closely by Jill, the QC who inspected that area; Lance, and two DCS employees; and finally John and another maintenance worker. The stakes have been raised from bickering between the sanitation and maintenance workers over who is responsible for what to the possibility that one or several NRs will be issued for anything ranging from hydraulic oil on an edible-meat conveyor belt to a piece of fat on the underside of a guardrail to chipped paint on a wall. Because Jill, as QC, is required to sign papers indicating that she has inspected the entire area, she is now responsible for any deficiencies in it. Sanitation and maintenance will also bear some of the blame, but the bulk of the anger from Roger and Bill over any pre-operational NRs will be directed at the quality-control office. Knowing which areas have been thoroughly inspected and which have been relatively neglected, Jill now tries to deflect Donald's attention away from the areas that are more likely to result in an NR.

In these inspections time works against the QC and in favor of the USDA inspector. Out of a list of hundreds of areas and equipment on the kill floor, the inspector will randomly select twelve, three in each of the four zones. The inspector has about forty-five minutes to go over these twelve locations—an average of three minutes and forty-five seconds per area. This amount of time alone makes it likely that the inspector will find something that the quality-control worker has missed. In addition, the inspector can single out anything that happens to catch his eye while walking to and from the preselected spots. Such random finds are referred to in the NRs as "stumble-on" deficiencies. And Donald is extremely good at finding problems. No matter how thoroughly Jill and I have inspected a particular area or a piece of equipment, Donald always seems to know exactly where to look or which surface to run his hand under in order to uncover something, sometimes minor, sometimes egregious, that we have missed.

Zone 4, the gut room, is a particular nightmare. Chunks

of fat and tissue inevitably escape even the best efforts of the sanitation crew and the QCs, only to be found by Donald's keen eyes and touch. Indeed, the question is never really whether Donald will find a problem: he nearly always does. It is whether Donald will choose to write up what he finds in an NR or whether he will simply tell us to get it cleaned and move on. Technically, of course, Donald can issue an NR for any contamination he finds since the pre-operational Sanitation Standard Operating Procedure (SSOP) written by the plant and approved by the USDA under the HACCP system states that before the start of production, all surfaces and equipment on the kill floor must be sanitary and free of contamination. Following the standard to the letter of the law would mean dozens of NRs each time Donald performs a pre-operational inspection. Instead, Donald reports only certain deficiencies. He seems to have no systematic approach for what he chooses to report and what he lets slide; if there were a system, our interactions with him would be far less volatile and stressful. Jill tells me early on that Donald writes NRs only for contamination on surfaces that directly touch the cattle carcasses or the meat and issues verbal warnings for contamination on non-contact surfaces. Yet within my first week as a quality-control worker, Donald writes an NR for dust and grease on the top of the 185 cabinet, an area that never comes into contact with the carcasses, and issues a verbal warning for dried blood specks inside the funnel for the head meat.

Much, it seems to Jill and me, depends on Donald's mood. If it is a Monday morning and Donald has just returned from a pleasant weekend, then his inspections will fly by quickly, and he will spend more time talking with us than looking at the equipment. But on other mornings he may emerge from his office with a scowl on his face, giving only a cursory grunt in response to our greetings, and proceed to scour every square inch of the equipment on his list. I notice that a large part of Jill's work with Donald seems to consist of engaging him in conversation, reaching out and touching his arm at key moments to distract him, and laughing obsequiously at his jokes and asides, many of which carry sexual undertones.

Despite this surface conviviality, it soon became apparent that a fierce undercurrent of animosity existed between Jill and Donald. Whenever Donald peered down to look under a piece of equipment, the smile on Jill's face would disappear, sometimes changing into an outright sneer. Likewise, Donald often seemed to take a perverse delight in beginning an inspection in a lighthearted manner, leading both Jill and me to believe that we were going to have an easy morning, then, when the inspection was almost over, discovering something for which he "had" to write an NR.

One morning during my first week as a QC, Donald was inspecting the underside of the omasum washer in the gut room when the batteries in his flashlight gave out. Jill laughed, shrugging her shoulders at his inability to continue the inspection, despite the fact that both she and I had working flashlights clipped to our belts. Hoping to change the climate from one of competition and animosity to one of cooperation, I handed Donald my own flashlight. He thanked me and continued his inspection, finding a few things wrong but issuing only verbal instructions to "take care of" them rather than writing an NR. My gesture of cooperation seemed to have worked.

Back in the privacy of the quality-control office, however, Jill lashed out. "Why would you do a stupid thing like that?" she yelled. "If his flashlight doesn't work that's his problem, not ours. This guy is not on our side, and it's not our job to help him. Don't you ever do something like that again!"

I protested that my strategy paid off; we did not receive any NRs.

"God, Tim!" she seethed. "Do you think that just because we're nice to Donald he's going to be nice to us? How stupid can you get? Donald is not like that. He is going to take advantage of us any way he can."

Jill's message was constantly reinforced by Roger, Bill, and all the red-hat supervisors during my time as a QC: this was war, and there would be no aid or comfort offered to the enemy. I learned quickly that my performance as a QC was measured by whether we received NRs, not by the cleanliness of the plant or the healthiness of the meat. The threat of NRs became the primary horizon of my working day, beginning at five in the morning and not ending until five or six in the afternoon when we had completed our paperwork. If one of us began the day with an NR, Roger would call us all into his office and bark, "These NRs have got to stop." It was unambiguous: my performance as a QC hinged on my ability to "handle" Donald.

Roger and Bill Sloan, in turn, felt their own pressure over receiving NRs. In the front office, Katherine and Sally kept careful track of the number and types of NRs issued. Each report, by law, requires an official response from the company stating what corrective actions will be taken to rectify the problem specified in the report. In the case of pre-operational deficiencies, the official company response is almost always that the relevant sanitation workers will be retrained and sometimes that the QC responsible for inspecting that particular area will also be retrained. In practice, this retraining amounts to nothing more than a verbal warning to "be more careful" the next time, but on paper it gives the appearance that the company takes the deficiency seriously and is actively working to correct it.

After the pre-op inspection is completed, one QC will perform a variety of setup tasks on the kill floor while the other goes to the cooler and swabs the flanks, briskets, and rumps of eight randomly selected carcasses from the previous day's production.[1] Later in the day, the QC will also cut samples of the head, cheek, and weasand meat and send them, along with the refrigerated swabs from the morning, to an on-site lab housed in a trailer in the slaughterhouse parking lot for bacterial-level testing. All swabs and samples are collected with care for cross-contamination, with gloves changed after each swab and hooks and knives sterilized meticulously. The QC also photocopies the food-safety paperwork from the previous day and delivers it to Sally in the front office, where she reviews it for errors or inconsistencies that might generate an NR.[2]

Once the slaughtering begins, typically soon after 6:00 A.M., the bulk of the QC's day is structured around Critical Control Points, or "CCPs." These tests play a highly specific role in the Hazard Analysis and Critical Control Point food-safety inspection system initiated in meatpacking plants during the Clinton administration. The HACCP system is widely viewed as reducing direct control of the federal meat inspectors in the inspection process and redirecting their primary attention to the documentation produced by the slaughterhouse's own quality-control workers rather than to the actual production process itself. Donald often refers to HACCP as "Have a Cup of Coffee and Pray."

Within the HACCP food-safety structure, the value of the CCPs depends on a strategy of random sampling at key points in the production process as a measure of the overall safety of the entire food supply. Slaughterhouse management and the USDA have jointly determined that the kill floor HACCP program will involve three CCPs, each performed hourly: a total of twenty-four tests a day. The first (CCP-1) involves a random hourly check of carcasses after they leave the trim rail just before they enter the beef wash cabinet (see the "QC" near circle 91 on figure 2). A total of sixteen carcasses, eight at ground level and eight at an elevated height, are inspected each hour, beginning at a specific time determined by randomly generated numbers. To pass, each carcass must be completely free of any visible trace of fecal material, ingesta (straw), milk, grease, or oil, a standard known as zero tolerance. The second test (CCP-2) consists of using sight and touch to perform hourly inspections of approximately forty pieces each of weasand, head meat, and cheek meat for contaminants. The third test (CCP-3), performed immediately after each CCP-1, involves looking at the thermometer readings on the 185 cabinet. The results of each CCP test is recorded on a standard form, and these forms are inspected daily by Donald in the quality-control office.[3]

After my initial training period, Jill and I rotated on a weekly basis between the CCPs, with one of us responsible for CCP-1 and CCP-3 and the other for CCP-2. I learned quickly that each CCP had its own rhythm and set of risks. CCP-1 and CCP-3, which were covered together because of their close proximity on the kill floor, could be carried out fairly quickly. At the rate of one half-carcass every six seconds, thirty-two half-sides take just over three minutes to examine. To inspect the carcasses, I would stand on a hydraulic lift and use a curved metal hook anchored in a plastic base to rotate each carcass as it sped by me on the overhead rail. Jill would explain repeatedly how important it was to peer intently at the carcasses lest Donald be watching and issue an NR for inadequate CCP inspection. Although I knew from reading the zero tolerance regulations that I was looking especially for fecal material, milk, or ingesta, I had very little sense of what these contaminants actually looked like on a cattle carcass. But when I asked Jill to describe what I should be looking for, her retort was a short and sarcastic: "Don't you know what shit looks like?"

The only cattle feces I had seen before working at the slaughterhouse were big clumps of manure in fields or dried and caked on the hides of live cattle. Against the ghoulish white flesh of skinned cattle that have passed through a scorching lactic-acid wash, I had little notion of what, exactly, "shit looks like." The USDA defines fecal matter as green or yellow material of a fibrous texture; under the zero tolerance standards, the discovery of a speck as small as one-eighth of an inch by one-eighth of an inch anywhere on the carcass is sufficient to justify an NR. I also had to watch out for ingesta and milk. Ingesta is defined as straw that has been in the cattle's mouth or jowls at the time of slaughter and has been transferred to the carcass itself. Milk can be a potential contaminant if a cow's udder has been slit open during the skinning process. In addition to the fecal, milk, and ingesta trifecta (referred to on forms as "FMI"), grease and "rail dust" are also potential contaminants. Used by maintenance to keep moving parts functioning properly, grease and oil can drip onto the carcasses from the overhead rails if too much has been used or if a mechanical part has malfunctioned. Rail dust consists of small flakes of dried grease knocked off the overhead rails as the trolley wheels pass over them.

Spotting these contaminants as the carcasses move past at high speed is extremely difficult. The QC must make a rapid visual scan of the carcasses and sort between discolorations and small splotches that are a natural part of the carcass and anomalies that might signify contamination. Folds of fat and muscle give the carcasses an uneven texture and create small valleys of shadow even under the bright glare of the overhead halogen lights, which are especially concentrated over the CCP-1 stand. In addition, the lactic-acid prewash spray creates black and brown "burn" spots on the carcasses that look much like grease and rail dust. Explaining to a newcomer what a QC should look for cannot extend much beyond "Don't you know what shit looks like?" and the ability to pick out tiny, almost microscopic bits of contamination comes only through hard-won experience.

As Donald would often point out to Jill and me, although the media hypes "mad cow" disease, it is *Escherichia coli* (*E. coli*) transmitted through feces that is much more likely to cause food poisoning. Unfortunately, fecal material is the hardest contaminant to spot and identify. Black grease and rail dust can be seen on the pale carcasses; fecal matter blends in with the carcass and is often so tiny that the eyes can miss it even when the QC is looking directly at it.

One indicator of the difficulty of spotting these contaminants lies in the fact that by the time the half-sides reach the CCP-1 stand, they have passed through the final trim rail, where six line workers and two USDA line inspectors do nothing but look for and trim off these contaminants (circles 86–89 in figure 2). Even after ten pairs of trained eyes have scanned the tops and bottoms of these half-sides, the rapid speed of the carcasses on the line makes it possible for fecal, grease, and rail-dust contamination to slip by unnoticed.

If Jill or I do find contamination on one of the half-sides during CCP-1, the implicit and explicit expectation from the kill floor managers is that we will not document it. Instead, we are to write "No FMI" on the entry for that hour's inspection and orally notify one of the trim-rail red-hat supervisors about the contamination. He, in turn, will tell the employees on the trim rail to "be more careful." This, of course, constitutes a direct violation of the sampling strategy upon which the entire logic of the food-safety inspection system on the kill floor rests, potentially multiplying the negative effects of finding contamination far beyond a single carcass since the samples are meant to provide an indication of the cleanliness of the larger population of carcasses.

The *official* procedure after contamination is spotted on a CCP test calls for the immediate shutdown of the line, the removal of the contaminant from the carcass, and the recording of the carcass number and time of failed inspection on the CCP-1 documentation sheet. After that, a multistep process termed "Regaining Control of the Critical Control Point Failure" is initiated. The first step consists of a consultation with Bill and Roger Sloan and the red-hat supervisors to determine where in the production line the contamination was most likely to have originated. For example, if the contaminant is on the rump of the carcass, we might hypothesize that it originated on the dirty side of the kill floor, possibly with the tail ripper or the legger. Perhaps one or both of these workers has not been sanitizing his knife in between carcasses, causing the contamination. Although this is always a game of probability and blame assignment, it will be recorded with absolute certainty in the documentation sent to the USDA, worded something like this: "Fecal contamination on right rump area of carcass caused by improper knife sanitation by first legger."

In the second step, the management specifies what corrective actions have been taken. Like the corrective actions for NRs issued during pre-operational inspection, these often read, "Employee responsible for the contamination counseled in correct knife sanitation procedures," or "Employee retrained in correct knife sanitation procedures." The word *counseled* simply means that a verbal warning has been given to the employee, and no additional paperwork is required to substantiate this. The word *retrained*, however, means that the employee has to sign a sheet of paper verifying that he or she has indeed been retrained in the correct procedure. Typically, "retraining" is reserved for a second failure within a short span of time. If a third CCP-1 failure is linked to a particular employee soon after the first two, the corrective action documentation might state that the employee has been "disciplined," "suspended," or even "terminated."

Once the corrective action has been taken to rectify the alleged source of the contamination, the third step, "Re-Establishing Control," is instituted. This involves waiting until the first carcass that has passed the source of the alleged contamination after the corrective action has been taken moves through the line to the CCP-1 stand. When it gets there the QC performs an unscheduled test to verify that the system is once again under control. If all thirty-two sides examined in the unscheduled test show no contamination, then the HACCP system is considered back "under control." All carcasses between the last successful CCP-1 check and the unscheduled CCP-1 check must then be tracked down in the cooler and labeled for individual inspection before being released to the fabrication department. At a line speed of three hundred cattle per hour, the number of carcasses needing reinspection after a failed CCP-1 test is usually in the hundreds.

Because these hundreds of carcasses are already railed tightly together in the cooler, they cannot be reexamined until they are moved off the holding rails onto the overhead rail that conveys them from the cooler to the fabrication department early the following morning. Several workers, usually from the trim rail, are assigned to come in early and stand in the cooler as these carcasses pass by. Each carcass is then examined on the top and the bottom and trimmed for any contamination as it moves through the cooler. If any additional contamination is found, this, too, must be documented.

From the kill floor manager's point of view, the costs of documenting contamination during a CCP-1 test are high. Not only is it an admission of contamination in documents accessible to the public through the Freedom of Information Act, it also means stopping the production line, filling out numerous forms, and assigning valuable workers to reexamine several hundred carcasses.

Similar problems with documentation plague the CCP-2 tests of weasand, head, and cheek meat. Because this meat is routinely included in hamburger, the USDA prevailed over the objections of the plant in requiring a CCP test for them. Unlike the CCP-1 test, during which the QC stands in one place as the half-carcasses pass rapidly by, CCP-2 tests involve picking up a box of each product and inspecting a sample of approximately forty pieces of meat for signs of fecal material, milk, or ingesta. Ingesta presents the largest threat of contamination. Sometimes a piece of straw can be as long as three or four inches; more common, however, are smaller pieces, ranging from an eighth of an inch to an inch long. Under zero tolerance, ingesta of any size on the head, cheek, or weasand meat is sufficient for it to fail the CCP-2 test and initiate a sequence of actions analogous to those performed when a half-side fails a CCP-1 test. All boxes of the failed meat produced between the most recent passed inspection and the first successful inspection after the failed inspection are brought back from the freezer and reopened, and each piece of meat in the box is individually examined for contamination. This can involve as many as twelve to fifteen boxes; at sixty pounds a box, the QC has to perform a manual inspection of more than 720 pounds of meat. In addition to the lost time for the reinspection, the failed CCP-2 test also results in an NR.

The different amount of time it takes to conduct a CCP-1 versus a CCP-2 inspection gives Donald a much greater advantage in the CCP-2 inspections. If he chooses to join a QC on the CCP-1 stand, he has only a few seconds to inspect each carcass visually as it moves by. In the CCP-2 test, however, the meat is right in front of him in boxes that can be opened and inspected more carefully. Donald will often ask for a box that has just been passed by a QC to be brought back to the CCP-2 table, where he will meticulously examine each piece of meat, sometimes finding a piece of ingesta that the QC has missed. When this happens, he will issue an NR not only for the contamination itself but also for the failure of the QC to find it during the test.[4]

The threat of receiving a double NR at the CCP-2 test is exacerbated by the explicit expectation among management that the QC will *not* document contamination but instead will quietly notify one of the red-hat supervisors that ingesta has been found so that the problem can be corrected unofficially.

After I joined the quality-control department, a situation arose on two consecutive days that illustrates how these "unofficial" procedures operate in practice at the CCP-2 test station. On the first day, Jill was checking a box of cheek meat when Donald came to stand beside her. One of the pieces of meat Jill picked up had a piece of straw on it, and because Donald was standing right next to her, she had no choice but to document the ingesta and fail the box. Whenever Jill or I fail a box of meat, we are required to get the digital camera and take a picture of the straw laid next to a ruler; Roger and Bill want to be able to claim that Donald has exaggerated the size of the contaminants in his previous NRs by showing photographic evidence of how small they really are. After photographing the ingesta, Jill prepared to fill out the paperwork: she determined in consultation with Enrique, the red-hat supervisor in charge of the head area, that the "loss of control" occurred at the head-washing station and that the employee there should be counseled to be more careful when washing the heads.

In reality, as Enrique often pointed out privately to Jill and me, it is not possible, given the current line speed, for the head-washing employees to guarantee that each head will be completely free of straw. The head washers use high-pressure hoses, sticking the nozzles in the back of the head to flush straw and other ruminant matter out of the mouth as the heads pass by on their hooks at the rate of about one every twelve seconds. The presence or absence of straw on the cheek meat and on the weasand, Enrique would tell us, is simply a matter of chance. The irony is that the entire HACCP system is predicated on the assumption that a loss of control represents a departure from the status quo, in which everything is under control; when it comes to keeping ingesta off the cheek and weasand meat, however, the status quo itself is a lack of control.

The following day, Jill and I were examining a box of weasand meat when Donald stopped about four feet away and leaned against the wall of the maintenance shop to watch us. As I ran my hand along the third or fourth piece of meat, I felt an irregular protrusion on the slippery surface of the weasand. Looking down, I saw a piece of straw approximately an inch long. Since I was still a trainee, I glanced at Jill for guidance, but she continued looking at the weasand in her hands and offered no reaction. Looking at Donald out of the corner of my eye, I could not tell whether he had spotted the straw.

I decided to document the ingesta and set the piece of weasand down on the metal inspection table. Donald immediately walked over and asked, “What do you have there?” Jill exhaled loudly, swearing, “Damn it, Tim.” I pointed out the piece of straw to Jill and Donald, then picked up the clipboard and wrote under the test scheduled for that hour, “Ingesta discovered on weasand meat.” I measured and photographed the straw. My procedure was exactly the same as that of the previous day, with one important difference: Roger, Bill, and Jill did not perceive what I did as *necessary.*

When we were alone in the QC office, Jill exploded at me for failing the weasand, arguing that Donald had been too far away to see the piece of ingesta. “He wouldn’t have seen that piece of straw. You should have just taken it off or put that piece of weasand back in the box and buried it under the other weasand.” I argued that the situation was the same as hers the day before, when she failed the weasand because Donald was standing next to her when she found the straw. She retorted that she had to fail that box because Donald “was breathing down my neck.” I mentioned that if Donald had decided to check the box of weasand after we were done with it and found the ingesta, we would have received two NRs: one for failing zero tolerance and another for failing to find the ingesta during our own check.

“Don’t you see what I always do when I put the weasand back after I inspect it?” she demanded. “I put it down in the box and move it all around so that he can’t be sure which were the exact pieces I inspected.”

"But we would have gotten an NR for the straw and for an inadequate inspection if he had inspected it and found it."

"You've been here long enough to know by now what this job is all about. We have to try to pass the product no matter what, and to beat the inspectors to the punch whenever there is a problem."

Later that same day when I was performing another hourly CCP-2 test, Bill Sloan walked over and stood next to me. After several minutes of silence he said, "You know, Tim, nothing is ever 100 percent perfect around here. If you find something on the product the best thing to do is just to tell one of the supervisors about it so that they can fix the problem. You know, Jill's really good at her job because she knows how to move her hands when she's checking the product so that if there is anything on it, she can take it off before the inspectors have a chance to see it. And she knows how to play it really cool if the inspectors are nearby. You see how she tries to talk to them to distract them if they're watching her do an inspection?"

I said something about not being sure whether Donald had seen the piece of straw. To that Bill replied, "If Donald's right there and it's obvious he sees it then you don't really have a choice. We don't want you to go to jail or nothing for doing the wrong thing, but if he's standing a ways off you just have to play it cool. Just watch what Jill does, she'll show you how to do it. We can't keep getting these NRs." And with that he walked away.

The ongoing, deliberate practice of not reporting contamination found during CCP-1 and CCP-2 tests often sparked justifications from the QCs and kill floor managers. During one test, when both Jill and I were up on the CCP-1 stand and several of the carcasses passing by clearly had fecal material on them, Jill turned to me and explained, "You know, I couldn't do this if we were the last ones checking these carcasses. But it's the people in fabrication who are the last ones who check these carcasses, so if anything were to happen [that is, if someone were to become ill from eating the contaminated meat] it's really on them." When I asked her directly how she felt about lying in the paperwork we fill out, her first reaction was hostility: "What do you mean we lie?" After I pointed out several recent examples of contamination that we failed to document she became more resigned and shrugged her shoulders: "Well, it's our job, you know? If we reported this stuff how long do you think we could stay in our jobs? Bill and Roger, they don't understand what we have to go through to do this job."[5]

Repeated comments from both Bill and Roger Sloan indicate, however, that they understand the work of the quality-control department exceedingly well. On more than one occasion they will say, half-jokingly, "You guys are here to sign all the papers so that if anything happens we don't have to go to jail." Bill and Roger also repeatedly assert that "nothing is ever 100 percent perfect and we have to be willing to be flexible. We do the best we can, and we don't let things get out of control, but sometimes we have to let the little things slide and keep sight of the big picture." This reasoning is often accompanied by comments about the unrealistic nature of zero tolerance standards and how in the days before zero tolerance the fecal material had to be larger than a certain size in order to justify shutting down the line or initiating corrective action: "Besides, the meat is much cleaner now than it was in the past."

Mixed in with such comments are personal attacks on Donald himself. As the USDA inspector who writes the vast majority of NRs, Donald is the central focus of Roger, Bill, the red-hat supervisors, and the QCs. Sitting up in his second-floor office, Roger often tracks Donald's movements and radios us about them, and unless we are dealing with other specific responsibilities, Jill and I are expected to either follow Donald or head him off before he arrives at the area he was going to inspect. Likewise, Bill and the red-hat supervisors radio in to report not only Donald's location on the kill floor but also his facial expressions and general demeanor: "Watch out, everyone, it looks like Donald is in a really bad mood today"; "He reminds me of a spoiled child who can't have his way when he looks like that." When Donald and Dr. Green inspect different areas of the kill floor simultaneously, Roger or Bill might warn, "They're doing their famous one-two-punch routine today, guys. Stay on top of them."

Letting things slide on the CCP tests is necessary, Jill, Bill, and Roger imply, because Donald is unreasonable and unwilling to "work with" the plant, choosing to write NRs for little things that the management might have been allowed to handle in the past. The implication is that if Donald were willing to be more flexible, the plant would be more up-front in its documentation of CCP failures. The possibility that the plant's refusal to document self-discovered contamination might be fueling what is perceived as Donald's "inflexibility" is never voiced.

Jill and I would sometimes wonder whether Sally and Katherine, the front-office staff responsible for the food-safety paperwork, "knew" about the gap between what is documented and what we observe. During my first two months as QC, I went to the front office every afternoon at three for "training" with Sally, the quality-assurance manager. We would sit in the upstairs conference room and watch painfully dull videos on food safety and HACCP produced by the American Meat Institute and other industry lobbying groups, and we would discuss the slaughterhouse's individualized HACCP plan and supporting prerequisites such as the Sanitary Standard Operating Procedures (SSOP) and the Good Manufacturing Practices (GMP). Sally never gave any indication that she knew that the information and procedures being presented in the videos and publications are radically different from what took place on the kill floor. On the contrary, she appeared to believe that the food-safety and quality-control operations on the kill floor followed the procedures of the official training she was giving me.

Roger, Bill, and Jill considered my training with Sally a waste of time, necessary only because the front office demanded it. The real "training," as Jill was fond of pointing out, could come only from my observing her do her job and then trying it myself. Roger would be quick to remind me whenever I returned from the front office that whatever I was learning there applied to a world in which everything ran 100 percent perfectly, as it was supposed to, but that back here in the "real world," "our people" had to be flexible to deal with imperfections and unanticipated situations as they arose.

Sally would occasionally don a white frock and hard hat and venture onto the kill floor to conduct "audits," standing with a list of hundreds of items and checking to see whether the procedures were being followed correctly. Most of these audits dealt with whether line employees were sanitizing their equipment correctly after making cuts on each carcass, or with whether they were washing their hands after using the bathroom or when coming back from break. Whenever Sally was on the kill floor, radio communication between Roger and Bill and the red-hat supervisors would typically turn to comments like "Oh here we go again" and "Keep a close eye on her, guys."

One day, Sally was standing next to the CCP-1 stand,

where Jill had just finished her check for contamination on the line of carcasses and had moved on to examining sixteen specific carcasses to be sure that the spinal cord had been completely removed from each one. Although not technically a part of the CCP-1 test, the spinal cord check was introduced as a result of the "mad cow" scare and the recognition that spinal cord material can be a carrier of BSE. For the most part, Roger and Bill were concerned only that the spinal cord be completely removed in cattle aged thirty months or older; in younger cattle with less risk of BSE, Bill and Roger would prefer that the spinal cord was completely removed but were not adamant about it. In keeping with the unofficial procedure of not documenting transgressions in writing, if a carcass passed by that still had part of the spinal cord in it, Jill and I would verbally notify a red-hat supervisor but write down nothing.

As Jill inspected the spinal cords, Sally noticed a carcass with part of the cord still in it. The animal was not tagged as being thirty months or older, and the usual practice was to let it pass the inspection point undocumented. Sally, however, followed the official procedure outlined in the company's HACCP and GMP plans and pulled a nearby red knob that immediately shut down the entire overhead line in the slaughterhouse. Although it took less than a minute for one of the yellow-hat workers to remove the remaining spinal cord and the line to start up again, Roger and Bill were furious at her for shutting down the line and, worse, for creating a situation in which the kill floor was required to document that there had been a problem in the production process.

On another occasion, Sally and Rick, the kill floor safety coordinator, discovered during a routine audit that workers from both the dirty side and clean side were using the same knife-grinding machine to sharpen their knives without sterilizing the whetting stone after each knife. Rick radioed Roger: "Those employees can't sharpen their knives on the same knife sharpener as the gray hats do," referring to the potential for cross-contamination.

Roger's reply dripped with sarcasm: "Well, you know what, Rick? Down here we live in the real world, and in the real world you can't do everything like you would want to. This is the *real* world, Rick." While Rick and Sally continued to walk around the kill floor with their checklists, Bill came up to me and complained, "Rick is so fucking nitpicky about everything. If you're in here looking around for seven hours and you can't find at least fifty mistakes, you're an idiot. When you're on the line working, your brain shuts off after about twenty minutes and every minute after that feels like another ten hours went by, but then you look down at your watch and it's like—Fuck, only five minutes have passed? When you do work like this, there's just no way you can concentrate on what you're doing for an entire workday, and if you can there's something seriously wrong with you. You have to fucking switch your mind off to survive. And when you do that, you make a mistake, and then you're like, Fuck, man, did I just do that?"

The QCs operate in the gap between the front office's training manuals, videos, and regimes of written documentation and the actual practices on the kill floor, which inhabit a world governed by the dual imperatives of keeping the line running at the highest speed possible and preventing NRs, work that is generally identified as running interference against the USDA. Our observations of Sally's interactions with Roger and Bill gave Jill and me no confidence in her authority to change things on the kill floor. We were even more unsure of Katherine's views concerning official versus unofficial practices; during my interview she emphasized the need for familial loyalty, and she rarely came onto the kill floor. As the ones filling out and signing the documentation on a daily basis, Jill and I were already deeply implicated in the failure to report contamination. Any attempt to alert the front office to the practice would also result in our own exposure. These ambiguities, along with the various rationalizations for not reporting, kept both of us from exploiting the gap between the front office and the kill floor as a possible means of creating change.

Whatever the justifications circulating among kill floor managers and QCs, there is no lack of awareness among the line workers of the contaminants on the carcasses. Since it is ostensibly the line worker's responsibility to remove these contaminants or to keep them from getting on the carcasses to begin with, it would be dangerous for the workers to talk openly about contamination on the cattle, since it would imply that someone was not doing his job correctly or that the line was moving too quickly to do the job properly. There were, however, notable exceptions to this silence.

One day in early November, I was writing up the standard "No FMI" on the CCP-1 form when Miguel, the worker who pinned each carcass with a number and weight as it passed the scale, pointed to a large piece of fecal material on the back of one of the half-sides passing by. I frowned and shook my head as if to say, "Man, that is so bad"—the typical ritual when workers or supervisors see contamination on a carcass. This frowning and shaking of the head conveys that the person who sees the contamination does not approve of it while freeing him or her from responsibility for it. Instead of simply shaking his head back, Miguel held up one of his hands and pantomimed with his other that I should document the fecal material. Then he motioned for me to come over to him and ordered, "Tomorrow you write that there is too much shit everywhere."

I stared at Miguel in surprise, and he smiled at me before beginning to laugh hysterically. "If you write that down, you come back the next day and you do this," he said, crossing his arms dejectedly in front of him and hanging his head. "I know," I replied; "if I write that down then the next day I come in and I am dead," and I made a slitting motion across my throat. Miguel laughed again, head thrown back. I too started to laugh a big, expansive, freeing laugh. Shielded somewhat by the moving line of half-carcasses, Miguel and I had found a space to mutually acknowledge the ludicrousness of the sustained pretense that the cattle were free of contamination. From that day on, Miguel and I shared a running joke. Whenever one of us saw contamination on the cattle, we would pantomime documenting it to the other person and exchange smiles and a chuckle.

Although CCP tests constitute the main reason for skirmishes between the USDA inspectors, kill floor managers, and QCs, there are other important issues that lead to recurring conflict. Most notable among these are the procedure concerning the tails and the problem of condensation. Tails are "harvested" from the carcasses by a worker standing on an elevated platform above the gut table while the carcasses are being simultaneously eviscerated. When the bung droppers cut out the anus and cap the end of the large intestine with a plastic bag, small flecks of fecal material are often transferred to the tails. In addition to severing the tails, the tail harvester is supposed to remove this fecal matter. The problem, however, is that the carcasses are moving by at such a high speed and the platform that the tail harvester stands on is so narrow that he has neither the time nor the room to do anything but cut the tails off and hang them on the line of moving hooks that transports them to the tail-packaging area.

Although Dr. Green, the USDA inspector in charge of the plant, outranks Donald, he is not as active, and does not spend as much time walking the kill floor and writing up NRs. He does, however, consider fecal matter on the tails especially worthy of attention, and almost every NR he issues is about this problem. On several occasions, Dr. Green would call me over to point out that the worker harvesting the tails is not dipping his knife into the hot water provided to sanitize it after cutting off each tail, potentially transferring contamination from one tail to another.

Whenever Dr. Green brings my attention to this issue, I have to make a big show of calling over one of the red- or yellow-hat supervisors and repeating the charge in Dr. Green's presence. The supervisor will then shake his head solemnly, agree that it is an egregious and unacceptable problem, and rush over to the tail-harvesting platform, climb the steps to the top, and passionately "counsel" the tail harvester about the need to sanitize his knife after every cut. The supervisor will then motion for the tail harvester to step back while he demonstrates exactly how to sanitize a knife. It is just possible, for the five or six carcasses that pass by during this farce, for the supervisor to maintain the superhuman pace and dexterity needed to sever and trim the tails, sanitize the knife, and keep up with the carcasses as they whiz by.

For his part, the tail harvester, who of course knows better than anyone that it is impossible to continue working at the pace being demonstrated by the supervisor even for fifteen minutes, much less for a nine-hour day, will nod his head in agreement, then bow it in apparent contrition. The supervisor, glancing down at me and Dr. Green, will give a big "thumbs up" to indicate that the problem is resolved, rush back down the stairs, and walk over to us, saying something like "Okay, now he knows." Dr. Green will give a curt nod and reply, "I'm going to let you guys go on this one because you were able to get the situation under control, but I'm trusting you to stay on top of this problem." The QC and the supervisor then genuflect in unison, and assure Dr. Green that the problem is indeed taken care of and that they can be trusted to "stay on top of it" in the future.

What makes the entire ritual even more farcical is that everyone, not just the tail harvester, the supervisor, and QC but even Dr. Green, knows that it is simply not possible to sanitize a knife after each cut, given the line speed and confined platform space. The tail harvester makes valiant efforts to sanitize the knife as frequently as possible, especially when Dr. Green is watching, but inevitably during the unending monotony of a fifty-hour work week he will not always realize that he is being watched. And whenever Dr. Green catches the tail harvester making five or six cuts without sanitizing his knife, the entire script is played out again.

Dr. Green is also intensely concerned with the "tail wash," a perforated metal tub that, much like a washing machine, uses a circular spinning motion to clean the tails before they are packaged in boxes for storage and shipment. The idea behind the tail wash, from the plant's perspective, is to rid the tails of undesirable materials like hair. For Dr. Green, however, the tail wash represents an attempt by the plant to cover up the presence of fecal material while exacerbating the problem. He would frequently refer to the tail wash as a fecal bath, pointing out to the QCs several times a week that if a tail with a bit of fecal material were put in the wash with other tails, the contamination would spread to all the tails in the wash, no matter how much cleaner they might appear to the naked eye.

Once the tails have been put in the tail wash, of course, there is nothing that Dr. Green can do to prove that they are contaminated, so instead he stands next to the chain that brings the tails to the washer and inspects each tail closely for any signs of fecal material. Whenever he does this, Jill or I will be alerted by the kill floor managers via radio, and we'll rush over, stand next to him, and try to spot any fecal material on the tails before he does. If we see it first, we'll take the tail off the hook and have it trimmed, averting an NR. If Dr. Green spots it first, he will either immediately write out an NR or issue a verbal warning to the QC, offering to "let it go this time."

Much speculation circulates among the kill floor managers and supervisors about the reason for Dr. Green's obsession with the tails. A popular theory, espoused by Bill and endorsed by Jill, is that at some point in the distant past, Dr. Green lost an argument with the plant over some other issue and is bringing up the problems with the tails in retaliation. Another, popular with Roger and several of the red-hat supervisors, is that despite his lower position in the USDA hierarchy, Donald is more competent and a tougher inspector than Dr. Green. The procedures for tail harvesting and washing, however, are something that Dr. Green can actually understand. Rather than risk being usurped by Donald on more complicated matters, Dr. Green chooses to focus exclusively on the tails because he knows he will always be right when he makes an observation. A third theory is put forth by both Roger and Bill whenever they are angry with Dr. Green for issuing an NR. They will radio the supervisors that Dr. Green has a kinky interest in tails, evidenced by the way he runs his hands over them during his inspection. This strange attraction, they will claim, extends to some other items of offal, such as hearts and livers. "Just watch him when he's inspecting them," they will tell us on the radio; "watch how he likes to take off his gloves so he can feel them all over with his bare hands."

Condensation on the kill floor provides another arena for constant skirmishes between USDA inspectors and QCs. Depending on the relative humidity outside the plant, drops of water sometimes form on the ceiling and railings above the kill floor, dripping periodically down onto the plant floor. Because harmful bacteria thrive in standing water, the USDA inspector will automatically issue an NR and shut down the line whenever he notices condensation dripping directly onto carcasses. When a QC notices condensation, she or he must immediately alert maintenance and the kill floor managers. In some areas, such as at the top of the decline, large fans can be set up on the catwalk to try to dry the condensation before it drips. Near the 185 cabinet and the CCP-1 stand, little can be done except to wait for a scheduled break and then wipe down the ceiling with squeegees and sponges.

Sometimes, however, the condensation in these areas becomes so heavy that Bill and Roger will order a maintenance worker to climb up onto the catwalk and the top of the rail so he can wipe the ceilings while cattle carcasses are still passing underneath. With moving cattle carcasses on the chain below, walking on the rail is unsanitary and risks contaminating the carcasses with material from the soles of the maintenance worker's boots; if a USDA inspector sees someone up there, he can issue an NR not only for contamination of the carcasses but also for willful violation of sanitary and good manufacturing practices.

If these combined measures still do not stop the condensation from dripping, the next step is to keep a careful watch on the USDA inspectors. Any time Donald or Dr. Green ventures near the condensation, one of the QCs will stand by one of the red knobs that can stop the entire production line. If either inspector glances up or shines a flashlight toward the ceiling, the QC will pull the knob, stopping the line and avoiding an NR by claiming that the plant was already taking corrective action. Like the pre-operational inspections, the CCP tests, and the rituals with the tails, this maneuver is an enactment of a complex game of stealth, cunning, and deception that would be comical were it not for the gravity of what is at stake. As Jill tells me repeatedly during my time as a QC: "Our job is to beat the inspectors to the punch and keep the line running at the same time."

From a theoretical perspective what is remarkable about this account of green-hat work in the midst of industrialized slaughter is not that massive violations of food-safety procedures take place on a routine basis. Nor is it the strategies of deception, falsification, and misdirection that are systematically deployed by kill floor managers, supervisors, and QCs to avoid detection of these violations by USDA inspectors. It is not the gaps between the formal procedures taught and outlined by the front office and the informal know-how that governs practices on the kill floor. Nor is it the common knowledge among line workers of the enormous gap between actual contamination and zero tolerance standards.

More remarkable than all these is the way a focus on food safety deflects the attention away from the work of killing onto the technical realm of hygiene. The possibility of perceiving and experiencing what happens in industrialized killing is diverted into elaborate performances and deceptions generated by the focus on food safety and the adversarial relationship between the slaughterhouse and the USDA. For the QCs, the experience of industrialized killing is refracted through a long chain of acronyms, tests, and statistics: HACCP, CCP-1, CCP-2, CCP-3, SSOP, GMP, lactic-acid concentrations, microbacterial carcass swabs, sterilization for head-meat sampling, pre-operational inspections, and corrective action plans. Sterile and banal in the abstract, the everyday practices of quality-control work on the kill floor transform these categories into fraught sites of power conflicts involving rapidly shifting alliances between the QCs and USDA inspectors, front-office personnel, subcontracted sanitation workers, kill floor managers, maintenance workers, supervisors, and line workers.

Unlike the white- and gray-hat line workers, whose workspaces are partitioned and whose divisions of labor are circumscribed, the QCs are practically the only participants in the work of industrialized killing who have access to, and are routinely expected to traverse, the entire kill floor. Despite this, the technical and bureaucratic requirements imposed by the skirmishes over their responsibilities for the control of quality serve to fragment, segregate, and neutralize the horror of the violent work being performed before their eyes just as effectively as walls and repetitive job functions do for the line workers. For the QC, sight and concealment work together and quarantine is possible even under conditions of total visibility.

VIII
Quality of Control

Job Number 121, Nonproduction Sanitation and Laundry Staff: maintains bathroom and lunchroom areas and offices. Does laundry.

Taken literally as measuring, sampling, and testing for food safety, quality-control work also has a second face that looks as much inward to the control of worker and animal bodies on the kill floor as it does outward to the control of food quality. Armed with the rationale of food safety and the technology of radio communication that puts them in instant communication with kill floor managers, QCs are also deployed by the management to assist with the surveillance and control of bodies, both human and nonhuman, to enforce the discipline necessary for industrialized killing. In the course of their various duties, QCs freely travel through otherwise segregated workspaces in multiple production zones, making them ideal agents to watch, record, and report the actions of line workers and cattle alike.

My promotion to QC involves a clear shift in my place within the slaughterhouse hierarchy. I wear a distinctive green hard hat instead of a white or gray hat, a laundered uniform with my name sewn on it instead of a T-shirt. I have a radio that puts me in constant communication with the kill floor managers and supervisors, as well as access to official paperwork, much of which I am responsible for filling out. I move freely about the kill floor performing a variety of tasks, rather than repeating the same rote job in a single, circumscribed place. I can use the bathroom whenever I need to. And ironically, like the upper-level management and outside visitors whose privileged, bird's-eye view I once renounced, I now walk the catwalk high above the kill floor several times a day, secretly watching the line workers.

My ground-floor perspective as an entry-level worker has been replaced by a very different kind of knowledge, shaped by interactions with plant managers, kill floor supervisors, federal meat inspectors, and front-office workers. My new vantage point allows me to chart the kill floor's spatial and labor divisions in microscopic detail enabling me to produce the map of the kill floor and the descriptions of each job that appear in Chapter 3 (see fig. 2) and Appendix A. I also gain a sense of the kill floor's relation to the cooler, the fabrication department, and the front office.

This new line of vision allows me to see that the management and the federal meat inspectors are not the monolithic entity they had appeared to be when I was a line worker. Fractures, divisions, and ongoing debates exist, particularly between the inspectors and the kill floor managers, as well as between the kill floor managers and the front-office staff. The powerful too have their hidden agendas and backstage spaces, and my promotion to QC gives me access to arenas of which I would never have been aware had I remained an entry-level worker.[1]

Although I try to maintain my relationships with the white and gray hats, it is painfully clear that my move to QC imposes costs. People who know me well, such as Ramón and a few of the cooler workers, remain friendly and open, inviting me to sit with them when our lunch breaks coincide, but even with them a certain strangeness arises. In part because of our different schedules, Ramón and I stop driving to work together; in the lunchroom, he no longer jokes about how many carcasses have fallen on the cooler floor or how many wads of fat he has chucked at another worker. (A month and a half after I joined quality control, Ramón was moved to the dirty side in the second-hock vacuum position. Working on the frenetically paced overhead rail without the protection from constant surveillance afforded by the cold of the cooler, Ramón became increasingly exhausted and quit just before completing a full year of work on the kill floor. His knees and hands had become inflamed from constantly standing in one place performing the same repetitive motion. He would look instead, he told me, for work in construction, perhaps tiling, which he did before coming to the slaughterhouse.)

In the slaughterhouse, muted suspicion marks my interactions with other line workers, noticeable in the sudden lapses in conversation when I enter a room, the quick glances, and the exaggerated deference in the locker room and on the kill floor. I try hard to overcome this, but no matter what I do it is as clear as the hard hat on my head and my monogrammed blue uniform that I am no longer a fellow line worker.

As a QC, I now have the imperative of keeping the line running at all times. Nothing, not even an NR, causes a greater uproar among Bill, Roger, and the red-hat supervisors than the line's being shut down for any length of time, for any reason. Every second a line is shut down is time that the line workers are being paid for doing "nothing," precisely accounted for in a weekly spreadsheet that calculates total labor costs, pounds of meat produced per man hour, and the labor cost to kill, eviscerate, and split each cow. These spreadsheets, in turn, are summarized by week in yearly aggregates, an exacting metric, in dollars and cents, of whether the kill floor managers are maximizing the amount of meat produced for each hour of labor.

A system of colored overhead lights and buzzers mounted on both the dirty and clean sides of the kill floor allows managers to identify the location of a line shutdown immediately, and a computerized system controlled from the kill floor manager's office automatically logs the length and specific location of each shutdown. At the end of the week, Roger Sloan will post the total amount of downtime in each supervisor's area of responsibility, often with handwritten notes of admonition or praise.

But the line workers celebrate shutdowns. If these stretch beyond a few minutes, the workers will leave their stations and congregate in small groups around the kill floor, sitting on stairs and tables and leaning against pillars. From the buzzing of conversation, the laughter, and the smiles arises a palpable sense not just of the physical relief that follows a temporary reprieve from the monotony of cramped and repetitive body motions but also of moral victory, a collective realization that for all its monstrosity the system itself still malfunctions in ways that liberate the workers, however temporarily. Line shutdowns invert the prevailing order in a visible way: as a QC, I join Bill,

Roger, the red-hat supervisors, and the purple-hat maintenance workers in an anxious rush to get the line moving again while the white- and gray-hat line workers sit observing us and making comments to one another about our performance.

Halting a process that maintains its hypnotic effect through perpetual motion, line shutdowns sometimes also highlight the fantastical nature of industrialized slaughter. One morning, just before eight, one of the hydraulic hoses on the side puller broke, spraying oil on the nearby carcasses. Rushing to the scene, Jill and I marked the carcasses with yellow "re-inspect" cards before Donald could arrive to issue an NR. Alerted to the problem via a "May Day" call over the radio, maintenance workers descended on the side puller like a swarm of purple hornets, pulling out wrenches, hammers, and screwdrivers. The metallic, astringent smell of hydraulic oil cut against the organic smells of the kill floor. Roger and Bill Sloan stood with their backs near the wall dividing the dirty side from the head table, arms crossed over their chests, shaking their heads. Every few minutes, one of them spoke into the radio: "How long is it going to take? How much more time?"

Finally, after about eight minutes, Bill radioed the supervisors to give the workers their fifteen-minute morning break, about an hour earlier than the scheduled 9:00 A.M. break. Calling morning breaks at will allows the kill floor managers to reduce the cost of line shutdowns by putting downtime on the worker's time rather than production time. The same strategy applies to lunch breaks, which are sometimes called up to an hour before schedule.

After about ten minutes, it became clear that the hydraulic-oil problem was not going to be solved soon, and I wandered a few steps from the side puller, stopping in front of the backers' belt. Made of white rubber and studded with little protrusions to give the backers a better grip, the conveyor belt is about three feet wide by six feet long, and it moves forward at the same speed as the overhead rail that transports the cattle carcasses, allowing the backers to work while the carcasses move.

Even after the overhead rail was stopped, the backers' belt continued to move. Three stationary cattle carcasses were suspended overhead, their tongues hanging limply out of their open mouths and dragging on the backers' belt as it spun forward. The little white rubber protrusions on the belt caught the rough tongues and pulled them forward, lifting the three heads ever so slightly before gravity pulled them back into midline with the carcasses. Dulled by death, the cattle's six enormous eyes gleamed foggily with the reflection of the overhead lights. This ghoulish dance was repeated over and over during the shutdown, with each cycle, from the snag of the tongue to the pull forward to the snap back to midline, lasting about seven seconds. A faint line of blood soon began to appear on the belt, becoming darker and darker as the shutdown stretched on until it changed into a pool, and then, as the blood congealed, a miniature landscape complete with hills and valleys and plains.

Those three enormous creatures, half-stripped of their hides, heads swaying back and forth in unison, with dangling tongues and listless eyes, are just one of many startling spectacles created by line shutdowns, unplanned snapshots that invite the attention from the overwhelming, perpetual motion of the industrialized kill floor to the fantastically grotesque aesthetic of what is hidden in plain sight, passing by at a rate of twenty-five hundred carcasses each working day.

For the kill floor managers, labor control, not aesthetics, is what matters, and primary oversight for the "productivity" of workers is given to the red-hat supervisors, each responsible for a particular production zone and each the front line between Roger and Bill Sloan and a multifarious and unruly tangle of "labor" problems: the problem of urination and defecation, which threatens clocks and timetables; the problem of tending to sick children and dying parents, which threatens the seasonal highs and lows of production needs; the problem of unannounced returns to Oaxaca or Chihuahua, which leaves the kill floor desperate during the winter months; the problem of fatigue and inattention; the problem of deliberate sabotage; the problem of incompetence, real and feigned; the problem of language; the problem of jealousies and rivalries between workers; the problem of headaches, sore throats, muscle pain, skin rashes, and hangovers; the problem of tardiness; the problem of dull and broken knives; the problem of whether to assign blame to malfunctioning equipment or worker error.

Most of the time, the struggles over these and other issues remain below the surface, invisible to an outsider but detectable to someone familiar with the kill floor environment. A worker arrives three minutes late on the line and the next day his supervisor denies a request for an unscheduled bathroom break. A worker misses a piece of hide and finds his request for a newly sharpened knife ignored. A supervisor perceives that a worker is being disrespectful and the worker finds his long-standing request for a vacation denied or lost.

Sometimes these subterranean struggles erupt into the open. One day in September, Gil, the supervisor on the dirty side, issued a three-day disciplinary suspension to one of the trimmers who had been added to help with the problem of feces-caked cattle. As he related it to Roger over the radio and in person to me later, the trimmer had been "giving him attitude all week." Gil had instructed the trimmer, a slim Latino with a long face and droopy mustache, to move to a different spot on the line. The trimmer had refused to move, saying that he would not leave his spot until Gil found someone to replace him.

For Gil, the trimmer's response was an unacceptable display of autonomy, threatening his own authority. The trimmer was there to execute Gil's orders, not to question or second-guess them. Furious, Gil radioed Roger and gave him a brief summary of the situation, concluding, "I can't have employees giving me attitude like this."

Roger's response was "Do whatever it is you need to do, Gil." Thus authorized, Gil gave the man a three-day suspension, effective immediately. The suspension was especially harsh because it included Labor Day, one of the few paid holidays in the working year. But a worker must work on both the day before and the one after the holiday in order to receive holiday pay; the suspension, in effect, robbed the man not only of three days of regular pay but also of a day of holiday pay. As Gil explained, "Maybe three days off and the loss of holiday pay will make him think twice about crossing me again."

The red-hat supervisors also use compromise, favoritism, and feigned ignorance of minor infractions in return for tacit promises from workers not to commit major ones. One area governed by supervisors' discretion is the granting of unscheduled bathroom breaks, which require a utility worker to step in and temporarily take over a job while the line worker uses the bathroom.

These unscheduled bathroom breaks are a constant source of annoyance for Bill. One day, frustrated by the number of bathroom breaks taken by the women who worked the spinal cord vacuums, he radioed the utility worker relieving them and ordered, "You do not have to give those ladies any breaks. In fact, tomorrow I don't want you to give them any breaks—that's what the nine o'clock and two o'clock breaks are for." Bill's directive to the yellow hat effectively usurped the authority of the red hat responsible for this area, revealing a gap between the informal agreements between the red hat and the workers and the discipline demanded by the kill floor.

Quality-control workers occupy a separate division, hybridized somewhere between management and line worker and dependent for their job security on Bill and Roger Sloan. Giving these workers a mandate about "quality" that implicitly entails, but does not explicitly require, surveillance of both the line workers and their red-hat supervisors allows the kill floor managers to monitor the red hats as well as the line workers. Technically, QCs are not managerial staff; like line workers, they are paid by the hour and have no direct supervisory authority over any other workers. Unlike line workers, however, QCs are not supervised by red hats; they report directly to Bill and Roger and have direct contact with staff in the front office.

Implement sanitation is one of the areas where red-hat supervisors typically "compromise" with their workers to the displeasure of Bill and Roger. Official procedures require kill floor workers to dip their implement (most often a handheld knife but sometimes a larger tool, like the hydraulic hock cutter) into vats of water heated to 185 degrees as often as possible in order to minimize the possibility of fecal and bacterial cross-contamination. The number of carcasses that are permitted to pass each worker before the worker must dip his or her implement in the hot water is not specified precisely, but it should not exceed three or four carcasses at a workstation.

From a line worker's perspective, dipping a knife or hock

cutter into hot water after every one or two carcasses can change a tolerable nine-hour day into an excruciating one. The physical motions required to dip the implement in the hot water include taking one's eyes and attention off the next carcass, turning to reach the hot-water basin, lowering the knife or piece of equipment into the basin, and turning back to the approaching carcass. While minor when considered in isolation, these motions, over the course of several hours, add up to a considerable loss in the already microscopic "rest" period the worker has between carcasses. More important, transferring focus from the line of moving carcasses to the stationary washbasin destroys the hypnotic numbness that can be produced by concentration on the blur of constant motion from the constantly moving line of carcasses. This numbness can help the worker lose track of time, relieving the almost unbearable psychological discomfort of focusing all his or her attention on the repetitive monotony of the line work. Acknowledging the impossibility of keeping to the official standards, red-hat supervisors almost universally under-enforce implement sanitation.

Roger and Bill, however, expect the QCs to watch the workers and try to catch them letting several carcasses go by without sanitizing their equipment. Aware of this, the line workers change their behavior whenever they see a green hat in the vicinity, often exaggeratedly dipping their knives into the hot water whenever it is obvious that a QC is watching. Not wanting to incur the wrath of Roger or Bill, the red hats often tip off the line workers whenever a green hat or a USDA inspector is approaching, bringing one or two fingers up to the eyes to signal that someone is watching them.

In turn, Roger and Bill up the ante by making it clear that they expect to hear from QCs throughout the day about workers who are not sanitizing their knives or are otherwise failing to fulfill their job requirements. A QC who does not radio in at least several infractions a day will invariably receive radio communication from Roger or Bill asking why they are being so quiet and reminding them to report anything they see.

The result is a cat-and-mouse game between the green hats on the one side and the red, gray, and white hats on the other. Occasionally a worker will be so hypnotized by the line or so new to the job that even when a green hat stands right next to him he will not sanitize his knife. For the most part, however, catching the workers requires the green hats to hide behind pillars, pretending to look one way when really focusing in another direction, or to use the catwalk. It is indicative of the surveillance component inherent in quality-control work that in addition to Roger and Bill, only the two QCs are given access to the catwalk that hangs above the kill floor, from which they can observe the line workers with less chance of being detected. In order to meet the unstated quota of misdemeanors and prove that they are doing their job, the QCs must publicly radio the red hat in charge of the workers whom they catch. These communications often fall into familiar patterns.

"Gil, Gil, do you copy, Gil?"

"Go ahead."

"Gil, I'm up here on the catwalk and ten cattle have gone by your second legger without any knife sanitization."

"Okay, I'll go talk to him." The reluctance in Gil's voice is evident, but he has little choice because the radio communication is on a shared channel, and if that worker is caught not sanitizing his knife again that day, Roger and Bill will blame Gil for poor supervision.

The kill floor managers' preoccupation with controlling workers extended to the front office as well. In my regular afternoon training sessions with Sally, the topic we would spend the most time on when we were not watching a training video was strategies for monitoring the line workers to make sure they were complying with sanitation requirements. Sally would sometimes indicate that the problem was that the workers were not sufficiently trained to do the job correctly. To address this, she started a video training archive: she would come to the kill floor and videotape a worker performing his duties in the "correct" manner so that it could be shown to other workers who performed the same task. At other times, she would suggest that the problem was one of willful misconduct on the part of the workers. To remedy this, Sally wanted to install video cameras throughout the kill floor, for she recognized that she was unable to catch the workers doing anything wrong because their behavior changed whenever they saw her in the vicinity.

That QC work entails employee control was also communicated through the training videos Sally and I watched. Some of these videos, produced by the American Meat Institute, feature camera-averse animal-science professors dryly reading manuals about the principles of HACCP and expounding on the temperature tolerances of various harmful bacteria. Others, however, have a larger agenda. In one series, titled *Food Safety Zone*, an actor representing a cameraman follows two actors representing meat-plant employees as they go through the various routines of their day. In the first episode, "Personal Hygiene," a white male employee wakes up in the morning and gets ready for work. The cameraman follows the employee into the bathroom and films him scrubbing his upper body with soap as the narrator explains the correct way to take a shower before work (be sure to wash with hot water and antibacterial soap, paying special attention to areas that might cause cross-contamination, transferring fecal material from the human body to the food product) and the proper procedure for applying a Band-Aid to a skin lesion or sore (make sure the skin lesion is completely covered by the Band-Aid and that it is not oozing). The employee starts to sneeze as he is stepping out of the shower, and the narrator explains that any food-production employee who feels the slightest bit ill must "be responsible" and call in sick rather than risk passing the illness on to others via the food supply. (At this, I could barely suppress a laugh, given that line workers at the slaughterhouse are regularly suspended or fired for being absent, even when they are sick.)

Emphasizing the motto "Be Clean and Work Clean," the video then cuts to two employees preparing for work at the meatpacking plant. Correct hand washing is again emphasized and broken down into seven detailed steps. "Be especially careful about hand washing after a bowel movement at work," the narrator cheerfully instructs, "because fecal matter can be on your hands and under your fingernails when you touch the food." The importance of wearing all personal protective equipment (PPE), such as hairnets, beard nets, boots, and frocks, is also underscored.

The next episode in the series, "Basic Microbiology," opens with one of the two workers talking about how excited she is to learn about microbes because she recognizes that as a food-production worker she plays a key role in making sure that the public does not become ill as a result of a mistake she or a co-worker might make.

The demeanor of the food-production employees in the training videos creates a cognitive dissonance between the world of the video and that of the kill floor. The well-rested, well-fed actor and actress in the film jump out of bed with a smile at the thought of spending another bright and beautiful day working in the meatpacking plant. Whether in the shower, applying a Band-Aid before work, or washing their hands after a bowel movement at work (itself an unrealistic luxury given the crackdown on unscheduled bathroom visits), these workers have only one thing on their minds: the safety and well-being of the unnamed masses who will be consuming the food they are producing. What a joy, what a privilege, what an honor it is to be a food-production worker, executing this most blessed and critically important task of feeding the growling stomachs of the world.

Sally would take these training videos seriously, often using the remote control to pause the video in order to underscore or repeat a particular point. Although she seemed cognizant of the difficulties involved in checking on workers while they were showering or applying Band-Aids to their sores and skin lesions before work, she would instruct me to stand unobtrusively in the kill floor bathrooms during breaks and after lunch to monitor whether the employees were following the correct seven-step hand-washing procedures outlined in the video.

"You won't be able to go into the women's bathroom," she might acknowledge with a hint of regret, "but"—perking up—"Jill should be doing that!"

In addition to spying on kill floor workers in the bathroom, Sally would also instruct Jill and me to eyeball the employees' knives as the workers left the kill floor at the end of the day. If a knife was at all unclean, we were to tell the employee to return to the knife-washing station and correctly sanitize it before leaving. "A lot of these people just don't know how to follow correct sanitizing procedures," Sally would note. Workers clearly resented having to show their knives to Jill and me each day, and although some of the knives I saw had bits of hair and fat clustered near the base, most were clean. Making each worker take the knife out of the scabbard, present both sides, and dip it in a bucket of disinfectant felt much more like an imposition of managerial will for the sake of control than a rational procedure motivated by a desire to ensure the knives were clean. Soon after I became a QC, Sally also began demanding that the workers' boots and aprons be sprayed with a special foaming disinfectant before they left the kill floor. There were only two hoses for this job, one at the clean-side exit and one at the dirty-side exit: this resulted in long queues of workers anxious to leave at the end of an exhausting day. The knife inspection and foam wash are humiliating rituals, parting reminders of the control the kill floor management exerts over the body of each worker; as a QC, I am complicit in enforcing that control.

QCs are also deployed by the management to watch over maintenance workers. In addition to performing the various CCP tests, QC workers must constantly measure various pressure and temperature gauges throughout the kill floor. They must take hourly measurements of the lactic-acid concentration in the spray applied to the carcasses at two points on the kill floor: once while the carcasses are still whole, just before their heads are severed (near circle 54 on figure 2 in chapter 3) and once just before the carcasses turn the corner and move down the decline into the cooler (near circle 97). Used to lower bacterial counts on carcasses, the lactic-acid spray, by USDA regulation, must maintain a concentration of between 1.0 and 4.5 percent with a tolerance of 0.5 percent. In order to minimize the number of failed bacterial-count lab tests, Roger and Bill Sloan insist that the acid-concentration level be kept as close to 5 percent as possible.

Testing the concentration of the acid in the mixing tank

involves drawing one cubic centimeter of the lactic-acid solution into a test tube, then adding one drop of phenolphthalein indicator solution. Once the indicator solution is added, N sodium hydroxide is added to the test tube, one drop at a time, and the test tube is shaken after each additional drop until the color of the solution in the test tube changes from clear to pink. Each drop of N sodium hydroxide indicates 0.1 percent lactic acid concentration: if it takes ten drops to turn the solution pink, this indicates a concentration of 1 percent; if it takes forty-five drops, the concentration is 4.5 percent, and so on.

Although there is a digital-control system mounted on the wall of the mixing room that can ostensibly be used to control the acid-concentration levels, none of the maintenance workers knows how to calibrate it, and I am simply told by everyone, including the head maintenance supervisor, that "it doesn't work." Thus, instead of using this digital-control system, Jill and I have to radio for a maintenance worker whenever the acid level needs to be changed, and one of them will come and manually adjust a screw in order to increase or decrease the amount of concentrated acid being added to the mixing tank.

Using a radio to call a maintenance worker to adjust the lactic-acid level, however, also alerts Bill and Roger Sloan, who monitor all radio traffic, that there is a problem with the levels. One morning early in my time as a QC, I radioed Steven, an older maintenance man, to tell him that the level was too low and needed to be adjusted.

"What is the level at, Tim?" Roger cut in over the channel.

"Two percent," I replied truthfully.

"Two percent!" Roger exclaimed. "Steven, if you can't keep those levels up higher, then I don't know what the hell we pay you to be here for, and maybe you just need to quit."

Later, in the quality-control office, Jill warned me not to call maintenance on the radio when there was a problem with the acid level: "You see what happened to Steven today," she said. "Just go and find them to tell them or just adjust it yourself."

Maintenance workers often err on the side of putting too much acid into the mixing tank, resulting in a concentration that exceeds 5 percent and sometimes is as high as 6 or 7 percent. Jill tells me during our training never to record numbers for the lactic-acid tests that are outside of the acceptable levels but simply to write down an acceptable number, tell maintenance to adjust the acid level, and wait for it to come within an acceptable range during the next test. She also tells me never to record that lactic-acid tests are lower than 4 percent because Roger and Bill will become angry; instead, I am to write "4 percent" and find a maintenance worker to increase the acid level.

After I had worked at the quality-control job for several weeks, Jill showed me how she finesses the situation when Donald or another USDA inspector asks to see a test being performed, which happens whenever they suspect that the acid levels are too high, either because of the number of blackish acid burns that appear on the cattle after they pass through the wash or because the mist from the wash irritates their eyes as they walk by the wash station. Because an acid-level test that exceeds the acceptable limit of 5 percent can lead to an NR, Jill keeps two different bottles of N sodium hydroxide in the lactic-acid mixing room. The first bottle, with a faded label, has its original nozzle; this is the bottle Jill uses whenever she wants to know the true level of concentration. Jill has cut down the nozzle on the second bottle of N sodium hydroxide; to the naked eye there is no difference between the nozzles on the two bottles, but the second bottle, with its slightly larger opening, releases more solution per drop than the first bottle. Thus, whenever Donald or another USDA inspector asks Jill to perform a lactic acid concentration test for him, she uses the bottle with the larger nozzle opening to reduce the number of drops needed to turn the solution pink.

Jill relayed this information to me as one quality-control worker showing another the tricks of the trade. She gave no indication that she thought she was doing something wrong, nor did she seem to consider what falsifying the concentration levels might mean for food safety. Rather, the issue was how to navigate between the conflicting pressures from kill floor managers who expected her to spy on other workers, maintenance workers whom she wanted to protect, and USDA inspectors who, from her perspective, were out to catch her in an error so that they could issue an NR. Charged explicitly with controlling food quality, green hats also work to enhance the quality of control, participating actively in Michel Foucault's "apparatus of total and circulating mistrust," in which "the perfected form of surveillance consists in a summation of malveillance."[2]

In addition to controlling human bodies, QCs are tasked with certifying the satisfactory control over the nonhuman bodies on the kill floor. They conduct a weekly internal audit of the treatment of the animals, and the paperwork from this audit is presented to meat purchasers to certify that the treatment of animals is being monitored in the plant. These meat purchasers in turn use the certifications to reassure their customers that their meat is humanely handled and slaughtered.

The humane-handling audit requires five separate forms. On the first, the QC must specify the rate of electric cattle prod use by the chute workers as the cattle are driven up the serpentine connecting the holding pens to the knocking box. This form specifies that using the electric prod on five or fewer cattle per hundred constitutes an "acceptable" use. Underneath these specifications are the numbers 1 through 100. The QC stands in the chute area and circles the number of every animal that passes by for which the electric prod has not been used and marks an "X" over the number of any animal that has been shocked as it passes by.

From my four days as a chute worker, I know that the actual use of cattle prods is about one shock for every three or four cattle, and that the chute workers, who have been made aware of the "acceptable" standards for electric prod use through an introductory training video, immediately modify their use of the electric prod whenever a QC or a USDA inspector enters the chute area. Nonetheless, even with this adjusted behavior, the actual number of shocks delivered ranges from ten to twenty per hundred cows. As with all of the other written documentation, however, the QC is expected to record results that fall within the "acceptable" range specified on the form.

The second form in the animal-handling audit concerns animal vocalizations. As with the first form, the quality-control worker circles the numbers of cows that moo or bellow as they walk up the chute; a vocalization of less than 1 percent of the cattle observed is considered acceptable. This moment in the operation of the slaughterhouse is unique in its attention, however cursory or disingenuous, to the actual, physical voice of the cattle. For the ten or fifteen minutes per week that it takes for a hundred cattle to file past the QC standing in the chute area, a human being is consciously and actively listening for their voices and interpreting those voices as an intentional communication of pain and suffering.[3]

The third form in the audit focuses on how many of the

cattle slip or fall in the chute and pen area as they are driven into the knocking box. The documentation procedure is identical to that of the first two forms. Because all three of these criteria specify activities taking place in the chute area, the QC often completes them simultaneously by observing the three variables for the same hundred cattle.

The fourth form concerns whether the knocking gun renders the cattle insensible after a single shot. The QC stands behind the knocker at the knocking box and, for a sample of one hundred cattle, circles the numbers of the cows that are knocked out by a single shot from the captive bolt gun and places an "X" over the numbers of the cows that remain conscious after the first shot. The "acceptable" range specified by the form is no more than three out of every hundred cattle. The number of shots it takes for a knocker to kill a cow varies, depending on the skill of the knocker, the working condition of the stun gun, the amount of air pressure being supplied to power the gun, and how much the cow struggles inside the knocking box. While performing this audit, I observe that some cattle need to be shot as many as four times. Far more common, however, is a shot that glances off the cow's head or does not penetrate deep enough, requiring a second shot in order to knock the cow unconscious. As with all other forms of documentation on the kill floor, the QC is expected to record results that fall within the "acceptable" range.

On one occasion, Sally decided to perform her own informal audit of the knocker. Armed with several photocopied diagrams with "X" marks illustrating where cattle should be shot with the captive bolt gun, she stood behind the knocker and began making notations on her clipboard. Immediately, Gil motioned me over to him. "What the hell does she think she's doing now?"

After observing for about ten minutes, Sally walked over to us and pointing to one of the diagrams on her sheet said, "I don't think they are shooting the cows in the right place. They should be shooting them a little higher on their heads." Gil responded by rolling his eyes; he had been working on the kill floor for almost twenty years, he noted, and the knocker was doing a fine job of shooting the cattle. I said nothing. Sally continued to point to the diagrams she was holding, trying to get Gil to look at them. Finally, Gil told her, "Look, if you have a problem with it, take it up with Roger or Bill," and walked away.

The last form in the audit requires the QC to stand in the area in front of the sticker platform to determine whether the cattle have remained unconscious as they hang upside-down on the chain. According to a training video I have seen in the front office, if the free hind leg of the cow kicks as it hangs on the rail, this is typically an involuntary muscle reaction and should not be construed as indicating that the cow is sensible. On the other hand, if the cow blinks in response to stimulus (such as waving a pen or snapping one's fingers near its eyes) or if it makes coordinated attempts to right itself by swinging upward, then these signals can be taken to indicate sensibility.

Jill and I took turns performing the animal-handling audit, and it was not long before I noticed that Jill's forms for the week were filled out before she went out to the chutes and knocking box to observe the cattle. When I asked her about this, her response was, "Nobody looks at these forms anyway, and we have to record what is acceptable whether it actually is or not, so why does it matter? Besides, it makes me sad to go out there and watch them get killed."

The animal-handling audit creates a paradoxical situation with regard to the individual animals. On the one hand, it is just that: an audit, broken down into five quantifiable components and designed, like so much of the documentation on the kill floor, to present a certain picture of industrialized killing—in this case of industrialized killing as humane. On the other, the result of the audit is to transform a physical confrontation with the killing of live creatures into a technical process with precise measurements of when the procedure counts as humane and ethical and when it does not. The inspector is looking directly at the animals; he or she is listening to their voices, but they are seen and heard only as criteria within a technical process, as data input. This technical dissociation operates for work performed outside of the audit as well, even to such extraordinary occurrences as an event that took place one day just before 11:00 A.M.:

"Gil, do you have a copy, Gil?" The voice comes crackling over my radio, and I recognize it as that of John Sloan, who supervises the unloading of live cattle in the pen area.

"Yeah, go ahead," Gil replies.

"Gil, I'm out here right now with the USDA and we have a little problem. One of our cows that's supposed to die right now gave birth in the pens to a calf, and the USDA won't let us have the cow until the afterbirth passes, so I'm just letting you know that the cow won't come in to die when it's supposed to die." I know from my time in the chutes that John means that the cow won't be killed with its designated lot, which is already being driven through the chutes to the knocking box.

Bill Sloan, John's older brother and the second-in-command on the kill floor, interjects in an irate voice, "Well, can't you just reach your hand in there and pull the afterbirth out?"

There is a silence on the radio before John replies, "Negative, Bill, the government's out here and they're not going to let us mess with this cow until it passes its afterbirth."

"Well, are you sure they're going to let us have it at all?" Bill asks.

"Yeah, ten-four, Bill," John replies after another silence. "We might have to get dirty with them, but they're going to let us have it. I'll let you guys know when it comes in to die."

As it turns out, the cow is the last "to come in to die," number 2,452 slaughtered that day. There is no mention over the radio of the fate of its newborn calf, at most a momentary nuisance within a process that views the cattle as the raw material inputs required to produce the desired output. The potentially powerful juxtaposition of a birth taking place in the midst of the work of killing is transformed into a technical dispute with the USDA over whether and when the cow can be slaughtered and further neutralized by the language typical of conversations about cattle in the slaughterhouse: the cow will "come in to die" rather than be killed. (Similarly, live cattle in the chutes are referred to as "beef," as in "Hey, guys, that beef has fallen down in the pens.")

In this view, the quality-control animal-handling audit extends, rather than challenges, the control over the cattle's bodies as a necessary condition for the work of industrialized killing. From the perspective of inputs in the production process, watching for balks, slips, and falls and listening for bellows and moos represents not attention to the suffering of the animals but rather awareness of potential disruptions in the steady supply of raw material. Concerns for the humane treatment of the animals and the regimes of documentation and observation created to attend to those concerns meld seamlessly with a production-centered view that maximizes the steady flow of raw materials onto the kill floor, one in which the animals are already beef even before they have been shot or bled.

And yet, as Jill's active avoidance of the observational component of the animal-handling audit suggests, frameworks that require the QCs to pay attention to the movements and vocalizations of individual creatures risk disrupting this control, creating moments in which human reactions to the animals' demeanor cannot be contained within the audit's technical criteria. There is no line for Jill's "It makes me sad to go out there" on the five standardized audit forms. Ironically, however, her response—to fill out the audit forms without making the actual observations—has the effect of further entrenching the technical framework that demands her attention to those animal bodies in the first place.

The work of quality control is as much about the quality of control over animal bodies, both human and nonhuman, as it is about the control of food quality. This control is sustained on the kill floor by a system of authority in which no one feels immune from being monitored by someone who has different incentives in the overall operation. Although the power relationships within this overlapping hierarchy of authority are differential, they are not absolute: red hats supervise line workers, but are overseen by QCs reporting directly to the kill floor managers, who in turn must answer to weekly spreadsheets that determine in precise monetary amounts how effective they have been in maximizing the pounds of meat produced per hour of labor. Accounts of power relationships in which one person or group controls another are insufficient to explain the complexity of the kill floor, where it is precisely the tension between the isolation of tasks and the ubiquity of surveillance that enables the work of massive, industrialized killing.

The QC's freedom of movement on the kill floor and

literal positioning from above fails to produce an attendant experiential understanding of the overall work of killing, demonstrating that surveillance remains compatible with compartmentalization and fragmentation. It is the explicit charge of overseeing and disciplining the bodies of workers and animals that allows the QC to look down on the work of killing from the catwalk, and yet it is simultaneously this work of surveillance that deflects the QC's attention from an experiential grasp of the underlying work of killing that is taking place, despite a view from above which renders each step of the killing process visually accessible. The QC looks at workers but sees failures to sanitize knives. The QC looks at and listens to cattle, but sees statistics on slips, falls, and vocalizations—quantifiable data points within a technical procedure designed to facilitate rather than confront the work of killing. Navigating the fraught demands of constant hierarchical surveillance over human and nonhuman bodies, the QC becomes an exemplary instance of how experiential compartmentalization is produced even, and perhaps especially, under conditions of total visibility.

IX
A Politics of Sight

A fear haunted the latter part of the eighteenth century: the fear of darkened spaces, of the pall of gloom which prevents the full visibility of things, men and truths. It sought to break up the patches of darkness that blocked the light, eliminate the shadowy areas of society, demolish unlit chambers where arbitrary political acts, monarchical caprice, religious superstitions, tyrannical and priestly plots, epidemics and the illusions of ignorance were fomented.
—Michel Foucault

"We've been watching you," Donald says to me suddenly one day. It is 6:00 A.M., and I have just completed the pre-operational inspection. The two of us are standing near the vacuum stands, out of sight of the kill floor manager's office. "We've been watching you," he repeats, "and we think you're a pretty good guy."

"Okay," I respond warily.

"Well, you know there's shit on this meat, don't you? We'd like you to talk to us about what's going on at this plant."

I am silent.

"You have kids, right? You want them eating this meat? Think about it, will you? Why don't you meet me tonight at nine at Dave's Pub and we can talk about it more then."

The rest of the day blurs by. I am three months into my quality-control position and already deeply uncomfortable with the simultaneous concealment (of food safety and humane handling violations) and surveillance (of kill floor workers) required by the job. Although I had initially planned to work on the kill floor for up to twelve months, my movement from cooler to chutes to quality control has already afforded me a thoroughness of access that I could not have anticipated when I first applied for work five months earlier. The initial fear that I might spend an entire year hanging livers has been replaced by physical, emotional, and psychological exhaustion from the grueling physical demands and ethical conflicts of quality-control work. Given the level of access already afforded by my three slaughterhouse jobs, both the rationale and my personal motivation for continued direct participant-observation work on the kill floor has been weakened. And now, from one of the head USDA inspectors in the plant, comes an invitation to become a whistleblower.

Meeting Donald at the bar that evening, I disclose that I am a researcher interested in writing an account of industrialized slaughter from the perspective of those who carry it out. Incredulous at first, he eventually accepts my explanation and reiterates that he would like me to consider testifying about what happens on the kill floor. I decline, noting that from the start my decision to access the kill floor as an entry-level worker without informing the management of my intention of writing about my experiences has included a commitment not to directly expose a specific slaughterhouse or individuals. At the same time, I offer Donald some insights into how the quality-control position in particular works, hoping to provide him with concrete steps he might take to increase the effectiveness of his food-safety monitoring. Several hours later, we part on amicable terms.

I quit at the end of the next day. During the day I let Ramón and a few others know that I am leaving, and we make plans to stay in contact. Noting the slaughterhouse's employment-at-will policy, which states, "Either you or the company may terminate employment at any time, with or without notice," I compose a brief letter of resignation to the kill floor manager and the human resources office, leaving a copy for each at the end of the work day. It states that I regret the abrupt nature of my resignation and lists the work equipment I have left behind in my locker: one employee identification card, one parking permit, four keys to various offices, two hard hats, one pair of leather boots, two pairs of rubber boots, one digital thermometer, one stopwatch, one black permanent ink marker, one flashlight, two knives, one sharpening steel, one orange hook, one plastic scabbard, one pair of safety gloves, one radio, and all uniforms not currently being cleaned.[1]

The prosaic list belies the complexity of observing and participating in the massive, routinized work of killing, work that remains hidden from the majority of those who literally feed off such labor. It is a complexity that highlights the unexpected sympathy between concealment and surveillance in the social strategies that distance dirty, dangerous, and demeaning work such as this from those it benefits directly. What I have called a politics of sight—organized, concerted attempts to make visible what is hidden and to breach, literally or figuratively, zones of confinement in order to bring about social and political transformation—must be alert to this sympathy.

As a whole, the slaughterhouse functions as Georges Bataille's cursed and quarantined boat, described in the epigraph to Chapter 1: physically, linguistically, and socially isolated in a zone of confinement that is inaccessible to most of society.[2] By removing the methodological distance that typically separates researchers from the social worlds they study and undertaking direct participant-observation research within the slaughterhouse, I sought in this book to provide insight into what it means, from the perspective of the participants, to carry out the work of industrialized killing. The divisions of labor and space inside the slaughterhouse walls revealed by this insider perspective exemplified not only how *distance* and concealment segregate the slaughterhouse from society as a whole but also how *surveillance* and concealment sequester the participants from the work of killing within the walls of the slaughterhouse itself.

To bring this work directly before your eyes, I mapped the slaughterhouse's uncharted interior, examining its contours and layout. Here we discovered that the slaughterhouse is not a single place at all. Its internal divisions create physical, linguistic, and phenomenological walls that often feel every bit as rigid as those marking off the exterior of the slaughterhouse from the outside world. From this internal vantage point it makes little sense to talk about "the slaughterhouse" as if it were a single entity within which responsibility for the work of killing can be pinpointed as belonging to this or that individual or department.

We first encountered the slaughterhouse as it appeared to visitors, who entered through its front office, where buttoned-down, khaki-wearing workers comfortably seated in leather chairs typed away in front of flat-screen computer monitors while discussing cattle futures on hand-free telephone mouthpieces. We saw the room into which these visitors were ushered, and where they gathered around a small shaded window cut into the only opaque wall, the wall of steel.

Framed in the window, line after line of white-frocked, white-helmeted workers stood shoulder to shoulder pulling chunks of meat off large moving conveyors, cutting it up quickly with their knives, and then throwing it back onto the conveyors. Past the fabrication department lay a vast, sepulchral cooler filled with row after row of silent, still carcasses. Down a steep flight of steps more shaking, swinging carcasses rattled by on their way to the cooler.

After passing through the fabrication department and cooler, we arrived at a hot, humid place whose straightforward name, "kill floor," belied an astonishing intricacy of divisions in labor and space. Workers in white, gray, green, yellow, red, and purple hard hats dotted the floor, most performing a single, repetitive task. Bung cappers and belly rippers, backers and bungee-cord attachers, cattle drivers and codders, heart trimmers and head chislers, prestickers and pregutters, paunch pullers and pizzle removers, supply-room staff and spinal cord extractors, toenail clippers and tendon cutters, tripe packers and tail baggers, Whizard knife wielders and weasand removers—121 distinct jobs in all make up the entity known as the kill floor.

Next, I described my own entry into the slaughterhouse. After an anxiety-ridden application process, I gained employment on the kill floor, where I was issued the white hat and rubber steel-toed boots of an entry-level worker. From here I moved through three jobs, each with a radically different relation to the work of killing. My white-hat cooler work introduced me to the daily rhythms of killing from a distance, killing mediated by the monotony of hanging liver after liver descending from on high via an unending line of moving hooks. It was a work of killing in which the most vivid experiences were throwing fat at friends and strategizing against those cocky fiends, the liver packers. Here the struggle was against monotony rather than the live animal whose steaming liver I now held between green-gloved hands, poised to thrust it onto a hook to be chilled, packed, and exported to distant places.

Next I donned the gray hat of the chute worker. My stint in the chute was short but critical: it was here that I joined ranks with the mere 8 or so workers—out of a total workforce of more than 800—who confront the cattle as living beings. And it was here that I experienced the work of the knocker, that 1 of the 120 + 1, who delivered the blow that knocked each creature unconscious. I listened in as the knocker was mythologized, even by hardened chute workers who themselves indiscriminately prodded cattle with electric shocks, as *the* killer among the 800, his job *the* work of killing, among the 121 jobs on the kill floor. Yet even here, at the one point in the long chain of industrialized killing where the animals are at once sensible and insensible, conscious and not conscious, it was impossible to state categorically that there was a moment when the cattle were alive and a separate, distinct moment when they were dead.

Finally, I entered the world of the green-hat quality-control worker, moving up in authority within the plant hierarchy and winning with that increased elevation the freedom to venture through divisions of space and labor that had once proved impenetrable. In a single day, I traversed wide swaths of the slaughterhouse, now down in the basement measuring lactic-acid concentrations, now out in the chutes listening for the vocalizations of the cattle, now at the Critical Control Point looking for fecal material, now standing at the down puller watching the hides ripped from the soft white carcasses. I gained vertical mobility as well, tiptoeing across the catwalk high above the kill floor, watching for workers who failed to sanitize their knives.

But this visibility did not necessarily translate into a deeper appreciation for the totality of the work of killing. Yes, it allowed me to map the kill floor with a level of detail unfathomable from the vantage point of any single line worker. Yes, it allowed me to listen in on the kill floor managers talking to red-hat supervisors over the radio. And yes, it offered me access to managers in the front office and to USDA inspectors with whom I would not have otherwise exchanged even a wordless nod. But all these things were measured out, acronym by acronym, in a steady drip of technical requirements and bureaucratic categories until I was straining on my toes, barely able to keep my nose above a rising tide of HACCPs, NRs, CCP-1s, CCP-2s, CCP-3s, pre-operational inspections, lactic-acid concentrations, sterile-carcass swabs, yellow tagging cards, gauge readings, dentition verifications, and vocalization, slip-and-fall, sensibility-on-the-bleed-rail, and successful-stunning-with-one-shot audits. Occupying the lofty vantage point of the catwalk above the kill floor, I discovered the sympathy between surveillance and sequestration as mechanisms of power, actively participating in Foucault's "apparatus of total and circulating mistrust," discussed in Chapter 1, which relied centrally on an ideal of total visibility for its effectiveness. This ideal worked in close symbiosis with the continued segregation of the work of killing itself, demonstrating the capacity for surveillance and sight to reinforce, rather than subvert, distance and concealment.[3]

The zones of confinement that characterize contemporary practices of industrialized killing replicate one another, beginning with the division between the slaughterhouse and society at large, followed by the divisions of labor and space between different departments within the slaughterhouse, and reproduced yet again in minute intradepartmental divisions. These zones segregate the work of killing not only from the ordinary members of society but also at what might be expected to be the most explicitly violent site of all: the kill floor.

Let us now imagine, as an alternative, a world in which distance and concealment failed to operate, in which walls and checkpoints did not block sight, in which those who benefited from dirty, dangerous, and demeaning work had a visceral engagement with it, a world in which words explained rather than hid and in which those with legal, medical, scientific, and academic expertise immersed themselves in the lived experiences of those they claimed authority over. Imagine, that is, a world organized around the *removal,* rather than the creation, of physical, social, linguistic, and methodological distances.

In this world, each time the state put someone to death, there would be a national lottery. Five people, perhaps including you, would be randomly selected to carry out the killing. The first would deliver the news to the prisoner's family, driving down back roads of hot asphalt or walking up steep stairs to a cramped tenement apartment, where the messenger would explain to the prisoner's family that in the name of the citizens of the state, their daughter or son, sister or brother would be injected, electrocuted, hanged, or shot in a month's time. The second person selected would prepare the prisoner's last meal, the third the chemicals, electric cords, rope, or bullets. A fourth would unlock the cell and accompany the condemned prisoner to the killing room. Once the fifth had strapped the prisoner into place, all five would gather and perform the killing.

In this world, each time a citizen relied on his or her citizenship to provide a privilege denied to a noncitizen, whether evacuation in the last days before a genocide or preference in college admissions, that citizen would need to experience directly the life of a noncitizen. Perhaps the citizen would lose his or her place on the helicopter to the noncitizen, driving home the arbitrariness of decisions made on the basis of birthplace. Perhaps the citizen would have to leave the seminar-room discussions of immigration and spend the day laboring beside undocumented workers planting flowers on the manicured campus lawn, paid under the table by the landscaping subcontractor who picked them up in front of Home Depot.

In this world there would be no "all-volunteer" armies, only a compulsory draft: first a selection of the sons and daughters of the decision makers and weapons manufacturers and then one organized by tax bracket, from highest to lowest. No dedicated spaces of "extraordinary rendition" would exist; the "enhanced interrogation techniques" performed on our behalf would be conducted before our eyes in our living rooms and public squares. No garbage truck would come in the dark morning hours of our dreams to take our waste out of sight and consciousness; the sick, the old, and the mad would not be shut away behind impenetrable walls of jargon and concrete; birth and death would not be locked up in institutionalized hallways. Sites of production would not be divorced from sites of consumption, and buying a pair of jeans would require the purchaser to touch the hands that sewed the seams. Every zone of privilege would exist in full contact with the zone of confinement that was its counterpart. In this world, the imperative that maps, proscriptively and prescriptively, the landscape of contemporary industrialized slaughter is reversed: to eat meat would be to know the killers, the killing, and the animals themselves.

The impulse to link sight and political transformation is strong. Returning to the earlier discussion of Foucault's articulation of the link between surveillance and power, a politics of sight that seeks to subvert physical, social, linguistic, and methodological distance in order to produce social and political change might be understood as a generalized Panopticon in which the prisoners have replaced the guards in the central tower that enables them to see without limits. The overseer's view, aimed at control and discipline, would be replaced with a view, accessible to all, aimed at transparency and transformation. Contrasting Jean-Jacques Rousseau's egalitarian vision with Bentham's disciplinary one, Foucault sketches the contours of this politics of sight:

> What in fact was the Rousseauist dream that motivated many of the revolutionaries? It was the dream of a transparent society, visible and legible in each of its parts, the dream of there no longer existing any zones of darkness, zones established by the privileges of royal power or the prerogatives of some corporation, zones of disorder. It was the dream that each individual, whatever position he occupied, might be able to see the whole of society, that men's hearts should communicate, their vision be unobstructed by obstacles. . . . This reign of 'opinion,' so often invoked at this time, represents a

> mode of operation through which power is exercised by virtue of the mere fact of things being known and people seen in some sort of immediate, collective, and anonymous gaze. A form of power whose main instance is that of opinion will refuse to tolerate areas of darkness. If Bentham's project aroused interest, this was because it provided a formula applicable to many domains, the formula of "power through transparency," subjection by "illumination."

If the disciplinary project of control underlying Bentham's Panopticon might be summed up as "each comrade becomes an overseer," the generalization of the Panopticon based on a society-wide dismantling of distance and concealment as mechanisms of power might be "each overseer becomes a comrade."[4] Take the same power of sight that serves the purposes of the dominating overseer in Bentham's Panopticon and use it as a counterforce: this is the strategy characterizing diverse movements across the political spectrum that seek to make visible what is hidden in zones of confinement as a catalyst for political and social transformation. It is a strategy that seeks to invert the "power through transparency" formula in the service of transformation rather than control and domination.

In her "ambiguous Utopia" *The Dispossessed* (1974), the science fiction novelist Ursula Le Guin juxtaposes a world that works through distancing with one in which the operating motive is transparency, offering a vision of the shape that such an inversion of "power through transparency" might take. Le Guin's protaganist, Shevek, is a brilliant physicist from the anarchist colony Anarres who is visiting Urras, the planet from which the Anarres anarchists seceded generations ago. Walking through a shopping district on Urras, Shevek is shocked and perplexed: "The strangest thing about the nightmare street was that none of the millions of things for sale were made there. They were only sold there. Where were the workshops, the factories, where were the farmers, the craftsmen, the miners, the weavers, the chemists, the carvers, the dyers, the designers, the machinists, where were the hands, the people who made? Out of sight, somewhere else. Behind walls. All the people in all the shops were either buyers or sellers. They had no relation to the things but that of possession."[5]

This relation of possession, characterized by concealment and distance of production from consumption, is reversed in the arrangement of space on Shevek's home planet, Anarres, where "nothing was hidden":

> The squares, the austere streets, the low buildings, the unwalled workyards, were charged with vitality and activity. As Shevek walked he was constantly aware of other people walking, working, talking, faces passing, voices calling, gossiping, singing, people alive, people doing things, people afoot. Workshops and factories fronted on squares or on their open yards, and their doors were open. He passed a glassworks, the workman dipping up a great molten blob as casually as a cook serves soup. Next to it was a busy yard where the foamstone was cast for construction. The gang foreman, a big woman in a smock white with dust, was supervising the pouring of a cast with a loud and splendid flow of language. After that came a small wire factory, a district laundry, a luthier's where musical instruments were made and repaired, the district

> small-goods distributory, a theater, a tile works. The activity going on in each place was fascinating, and mostly out in full view. Children were around, some involved in the work with the adults, some underfoot making mudpies, some busy with games in the street, one sitting perched up on the roof of the learning center with her nose deep in a book. The wiremaker had decorated the shopfront with patterns of vines worked in painted wire, cheerful and ornate. The blast of steam and conversation from the wide open doors of the laundry was overwhelming. No doors were locked, few shut. There were no disguises and no advertisements. It was all there, all the work, all the life of the city, open to the eye and to the hand.[6]

The contrast between Urras and Anarres captures perfectly the distinctions between visible/invisible, plain/hidden, and open/confined that, in theory, keep repugnant activities hidden and therefore make them tolerable. Breaching the zones of confinement and rendering the repugnant visible thus appears as an available tactic of social and political transformation. Le Guin's portrait of Anarres, a place where "it was all there, all the work, all the life of the city, open to the eye and to the hand," is powerful in its appeal. Le Guin is, in effect, inviting us to imagine a world in which physical, social, and linguistic mechanisms of distance and concealment are subverted; growing up on Anarres, Shevek is shocked by the way those mechanisms separate consumption and production on Urras.

But how might the work of killing fit into this society "where all is open to the eye and to the hand"? Would children be permitted to wander the kill floor, to work with, say, the lower belly ripper or make mud pies out of eviscerated livers? As part of this impulse for transparency, the food writer Michael Pollan advances the powerful idea of the glass abattoir, which he developed after visiting an open-air chicken slaughterhouse in Virginia:

> This is going to sound quixotic, but maybe all we need to do to redeem industrial animal agriculture in this country is to pass a law requiring that the steel and concrete walls of the CAFO's [Concentrated Animal Feeding Operations] and slaughterhouses be replaced with . . . glass. If there's any new "right" we need to establish, maybe it's this one: the right to look. No other country raises and slaughters its food animals quite as intensively or brutally as we [in the United States] do. Were the walls of our meat industry to become transparent, literally or even figuratively, we would not long continue to do it this way. Tail-docking and sow crates and beak-clipping would disappear overnight, and the days of slaughtering 400 head of cattle an hour would come to an end. For who could stand the sight?[7]

Like the open shop fronts and factories of Le Guin's Anarres, Pollan's glass-walled slaughterhouse is an attempt to counter distance and concealment as mechanisms of power by making all "open to the eye." The repugnant practices of the slaughterhouse (no other country slaughters its animals as brutally) continue only because they take place in a zone of confinement (the walls of the slaughterhouse), and these practices would come to a halt (disappear overnight) if there were a breach in the zone of confinement that made the repugnant visible (were the walls of the slaughterhouse to become transparent, literally or even figuratively). Reworded in this way, Pollan's glass-abattoir argument relies centrally on the assumption that simply making the repugnant visible is sufficient to generate a transformational politics: for who could stand the sight?

The rhetorical force of this question presumes a more or less standardized response, a generalized opinion, to industrialized slaughter made visible. Disgust, shock, pity, horror: the precise emotive label is less important than the assumption, the unarticulated expectation, of a reaction that would engender political action to end or transform the practices of industrialized killing. Pollan's glass abattoir is a powerful and concrete expression of the relations between "power through transparency" and "the reign of opinion" we encountered in Foucault's discussion of the Panopticon. These practices continue, Pollan implies, because they are hidden, shrouded in darkness, and confined to remote places. Under the light of everyone's gaze, under *our* gaze, they will wither and shrivel up, scorched by the heat of our disgust, our horror, our pity, and the political action these reactions engender.

Paradoxically, an assumption of "power through transparency" also motivates those who fight to keep the slaughterhouse and related repugnant practices quarantined and sequestered from sight. The recently proposed Iowa legislation (also under consideration in Florida) that seeks to criminalize those who make visible the hidden work of industrialized slaughter and other contemporary animal-production practices is also based on the assumption, shared by Le Guin and Pollan, that the act of making the hidden visible could generate political and social transformation. This legislation counteracts a politics of sight by seeking to create and maintain zones of concealment and areas of darkness around contemporary practices of food production. By criminalizing the production, possession, and distribution of records of such hidden work—where records are defined expansively to include "any printed, inscribed, visual, or audio information that is placed or stored on a tangible medium, and that may be accessed in a perceivable form, including but not limited to any paper or electronic format"(defined expansively enough, in other words, to include the book you are now reading)[8]—proponents of such legislation ironically underscore a key assumption of any politics of sight: the transformational potential inherent in making the hidden visible.

Pity (or horror, disgust, and shock), then, is the assumed response to slaughter made visible, both by those who seek to transform contemporary slaughter practices and by those who seek to maintain the status quo. In a politics of sight, pity and its related emotions carry the burden of transformation. Rousseau provided the clearest statement of the role of pity in social improvement: "Men would have never been anything but monsters if Nature had not given them pity in support of reason. . . . Indeed, what are Generosity, Clemency, Humanity, if not Pity applied to the weak, the guilty, or the species in general? It is pity which carries us without reflection to the assistance of those we see suffer; it is pity which, in the state of Nature, takes the place of Laws, morals, and virtue, with the advantage that no one is tempted to disobey its gentle voice. . . . It is, in a word, in this Natural sentiment rather than in subtle arguments that one has to seek the cause of the repugnance to evil-doing which every human being would feel even independent of his education."[9]

Against Rousseau's timeless universalization of pity and "repugnance to evil-doing" in the face of the physically and morally repugnant, however, we must set one of the central conclusions of Norbert Elias's *The Civilizing Process:* as violence is increasingly monopolized by the state and sanitized from the sphere of everyday life, we have redefined "the repugnant," expanding its frontiers and refining our response to it. In Elias's account, it is the increasing segregation and concealment of violence from the sphere of everyday life that leads to this expansion, a refinement and intensification of what Rousseau terms the sentiment of pity or commiseration, what Anthony Giddens characterizes as events that arouse existential questioning, what Hannah Arendt references as "the animal pity by which all normal men are afflicted in the presence of physical suffering," what Max Horkheimer terms "the solidarity of the living," and what Lev Tolstoy evokes in passages like the following: "When a man sees an animal dying, a horror comes over him. What he is himself—his essence, visibly before his eyes, perishes—ceases to exist."[10]

The expansion of the frontiers of repugnance as a complementary aspect of the distancing and concealment characteristic of the civilizing process is underscored if we contrast Tolstoy's universalized, timeless man, who reacts with horror at the sight of suffering, with accounts of institutionalized or socially sanctioned violence inflicted against animals in earlier times or in contemporary but more "primitive" societies. Take, for instance, the two descriptions offered below, the first separated from "civilization" by time and the second by space:

> In Paris during the sixteenth century it was one of the festive pleasures of Midsummer Day to burn alive one or two dozen cats. This ceremony was very famous. The populace assembled. Solemn music was played. Under a kind of scaffold an enormous pyre

was erected. Then a sack or basket containing the cats was hung from the scaffold. The sack or basket began to smolder. The cats fell into the fire and were burned to death, while the crowd reveled in their caterwauling.

Indifference to the pain of animals has been frequently observed among [contemporary] hunter-gatherers. Consider, for instance, the Gikwe Bushmen of the Kalahari Desert, a people known for their gentleness toward each other and toward outsiders. But this gentleness cannot apply, for obvious reasons, to the animals they have to kill for food. A certain callousness toward animal suffering is evident even when hunger is not a pressing question. Elizabeth Thomas, in her book *The Harmless People,* describes an event which, because it is quite ordinary, reveals a hunting people's deep, unreflexive attitude toward animal life. A man named Gai was about to roast a tortoise which belonged to his infant son Nhwakwe. Gai placed a burning stick against the tortoise's belly. The tortoise kicked, jerked its head, and urinated in profusion. The heat had the effect of parting the two hard plates on the shell of the belly, and Gai thrust his hand inside. While the tortoise struggled, Gai slit the belly with his knife and pulled out the intestines. "The tortoise by now had retreated part way into its shell, trying to hide there, gazing out from between its front knees. Gai reached the heart, which was still beating, and flipped it onto the ground, where it jerked violently." Meanwhile, the baby Nhwakwe came to sit by his father.

> "A tortoise is such a slow tough creature that its body can function although its heart is gone. Nhwakwe put his wrists to his forehead to imitate in a most charming manner the way in which the tortoise was trying to hide. Nhwakwe looked just like the tortoise."[11]

These accounts antagonize "civilized" sensibilities. They offend against that "horror" described by Tolstoy, the commiseration evoked in Rousseau, and the "solidarity of the living" addressed by Horkheimer: in short, they provoke reactions of physical and moral disgust in those whose frontiers of repugnance have expanded as a result of the operations of distance and concealment that we have recognized as the primary mechanisms of the civilizing process. Unlike Rousseau, who naturalizes pity as a "sentiment of Nature," Elias demonstrates that pity (like disgust, shock, and horror) is an emotive response that becomes increasingly refined and widespread as the frontiers of repugnance grow. These frontiers, in turn, expand in proportion to the advancement of a civilizing process that has as its central mechanism concealment and distance, the hiding away of what is distasteful. "Civilized" humans separated by time from the festival public killings of cats in sixteenth-century Europe or by space from the tortoise-eating Gai may react to these accounts with pity, disgust, and shock, but it is a reaction predicated on the operations that remove from sight, without actually eliminating, equally shocking practices required to sustain the orbit of their everyday lives. The work of killing detailed in this book is an account of precisely one such contemporary practice.

Here again, an unexpected sympathy between surveillance and sequestration is revealed, and seemingly contradictory ideas about the relations between power and sight are shown to be intimately connected in actual practice: both, it turns out, are modes of power capable of acting in concert to reinforce relations of domination. The very question "For who could stand the sight?" becomes historically intelligible only in the context of a "reign of opinion" dependent for its existence on the *continued* operation of distance and concealment, the *continued* hiding from sight of what is classified as repugnant. In this way, the ideal of the generalized Panopticon, of a world where all is open to the eye and the hand, and of a glass-walled slaughterhouse paradoxically relies on the very distance and concealment they seek to counteract for the emotive engine that is implicitly or explicitly assumed to generate their transformational power. The politics of sight feeds off the very mechanisms of distance and concealment it seeks to overcome; sight and sequestration exist symbiotically.

The answer to distance and concealment as mechanisms of domination, however, is not more distance and concealment. In a world characterized by the operation of physical, social, linguistic, and methodological distance and concealment as techniques of power, movements and organizations that seek to subvert or shorten this distance through a politics of sight are necessary and important. WikiLeaks, Transparency International, People for the Ethical Treatment of Animals, Operation Rescue, Human Rights Watch, Amnesty International, Doctors Without Borders, the Humane Society of the United States, the Humane Farming Association, Smile Train, the Open Society Institute—these are just a few of the vast number of movements that aim at the metaphorical equivalent of a world in which slaughterhouses are enclosed by walls of glass. Advancing dissimilar or even highly antagonistic political agendas, these movements nonetheless share a common politics of sight insofar as they deploy words, images, and social media to breach zones of confinement on the implicit or explicit assumption that once those breaches are created, a "reign of opinion" rooted in outrage, pity, disgust, sympathy, compassion, solidarity, shock, horror, or some other emotive response will lead to political action in the service of their desired goals. For who could stand the sight?

But as the demonstration of the potential for sequestration and sight to work in conjunction with each other suggests, it is a risky strategy and one that always yields imperfect results. "For photographs to accuse, and possibly to alter conduct, they must shock," writes Susan Sontag—to which we could add that if shock, like many other emotions, requires increasing stimuli to maintain itself, then we are not far from a strategy that demands increasing intensification of its representations of suffering, pain, and the repulsive in its effort to reduce their actual occurrences in the world. This intensification, in turn, would reduce the shock level of subsequent representations in yet another iteration of the symbiotic relation between sight and concealment.[12]

The account of the work of killing provided in this book suggests a much more nuanced relation between sight and sequestration than simple binaries between visible/invisible, plain/hidden, and open/confined can accommodate. Even when intended as a tactic of social and political transformation, the act of making the hidden visible may be equally likely to generate other, more effective ways of confining it. We have already seen, with the slaughterhouse quality-control worker, how isolation and sequestration are possible even under conditions of total visibility. A world where slaughterhouses are built with glass walls might lead in turn to one in which enterprising slaughterhouses charged people admission to witness or participate in repetitive killing on a massive scale. A world in which a lottery is used to select citizens to kill condemned prisoners might spawn a black market for the sale of winning lottery tickets, an opportunity to witness death up close under the sanction of the state. The logic of "who can stand the sight?" is as likely to be a basis for making a profit off the *pleasure* of feeling pity for the less fortunate as it is for the transformation of their plight. Making the repugnant visible, Sontag notes, may as well result in apathy as action: "The gruesome invites us to be either spectators or cowards, unable to look. Those with the stomach to look are playing a role authorized by many glorious depictions of suffering. Torment, a canonical subject in art, is often represented in painting as a spectacle, something being watched (or ignored) by other people. The implication is: no, it cannot be stopped—and the mingling of inattentive with attentive onlookers underscores this."[13]

Imagine once again a world organized around the removal of physical, social, linguistic, and methodological distances. Is such a world desirable? Is such a world possible? "An Ambiguous Utopia" is the subtitle of Ursula Le Guin's treatise on Anarres, that anarchist colony where all is "open to the eye and the hand." Insofar as the ideal of transparency exists in intimate relation with the mechanisms of distance and concealment it seeks to overcome, it too is deeply ambiguous. Concerned with the subjecting power of a generalized gaze, some will dismiss the ideal, like vision itself, as a trap. Others, placing their faith in a weight of opinion and the immutable timelessness of pity, will energetically advance the project of bringing every dark thing to light, demolishing every distance between what is seen and what is hidden. The ambiguity inherent in the ideal of transparency opens up a vast empirical research agenda that might incorporate instances of the politics of sight as diverse as the political movements that employ it today. The aim of such research would include a close specification of which conditions, contexts, and types of making visible are likely to be more politically transformative and which are likely to result in renewed forms of sequestration and concealment.

The account of industrialized slaughter provided in this book itself enacts a politics of sight, seeking to subvert particular physical, social, linguistic, and methodological distances separating the reader from the slaughterhouse. At the same time, it is also an account, from the perspective of lived experience, of how concealment and visibility are at work within the slaughterhouse, demonstrating that hierarchical surveillance and control are not incompatible with the compartmentalization and hiding from view of repulsive practices, even at the very site of killing. Where distance and concealment continue to operate as mechanisms of domination, a politics of sight that breaches zones of confinement may indeed be a critically important catalyst for political transformation. This politics of sight, however, must acknowledge the possibility that sequestration will continue even under conditions of total visibility. And, it must also remain alert to the ways in which distance and concealment provide the historical conditions of possibility for its effectiveness. These conclusions signal the need for a context-sensitive politics of sight that recognizes both the possibilities and pitfalls of organized, concerted attempts to make visible what is hidden and to breach, literally or figuratively, zones of confinement in order to bring about social and political transformation. In this book I have offered footholds for such a politics through a detailed account of how sequestration and surveillance distance the work of industrialized killing from society at large as well as from the very people who perform it. By means of these footholds, we might move toward a transformation not only of how the work of industrialized killing and analogous repugnant practices are seen but also of how, if at all, they are carried out.

Appendix A: Division of Labor on the Kill Floor

The numbers in this appendix correspond to the numbers within the circles on the kill floor maps provided in Chapter 3 (figs. 2–10) and should be read in conjunction with those figures. Bracketed numbers indicate how many workers perform a particular job; no number means that only one worker does the job.

1. *Cattle Driver* works in pens and squeeze pen; uses electric prods, paddles, whips, and voice to drive cattle into serpentine [4].

2. *Serpentine Cattle Driver* works in serpentine; uses electric prods, paddles, whips, and voice to drive cattle up serpentine and into knocking box. Also responsible for marking lot cow [3].

3. *Knocker* operates knocking box; uses air gun to drive captive-steel bolt into foreheads of cattle while they are suspended on center track and restrained by side walls.

4. *Shackler* shackles rear left hind leg of cattle with chain

suspended from overhead rail; can shackle either before or after knocked cattle fall onto green conveyor belt from which they are lifted up by the chain.

5. *Indexer/Hand Knocker* uses long metal pole to space cattle between "dogs" on overhead rail; uses cap-gun hand knocker to shoot cattle that show signs of sentience after passing through knocking box.

6. *Ear-Tag Recorder* uses paper forms to record ear-tag number and color of each cow; also keeps track of lot numbers.

7. *Presticker* uses hand knife to make incision along length of the cow's neck, giving the sticker access to jugular vein and carotid arteries. Must take care not to be kicked in face, arms, chest, neck, or abdominal area by cows that are reflexively kicking, or kicking because they have not been knocked completely unconscious.

8. *Sticker* reaches into the incision made by the presticker and uses hand knife to cut jugular veins and carotid arteries of cow.

9. *Tail Ripper* uses hydraulic scissor-type knife to cut off bottom third of tail; disposes of this down a chute; uses hand knife to make an incision from the anus to the teat or penis area.

10. *First Legger* uses hand knife to cut hide off rear right leg, opening hide to expose flesh underneath [2].

11. *Bung Capper* uses hand knife to cut around anus.

12. *First Hock Cutter* uses large hydraulic shears to sever right hind hoof about six to ten inches from end and deposits hoof in chute; uses hand knife to poke hole between tendon and lower leg bone.

13. *Belly Ripper* uses hand knife to make incision down length of cow, starting from udder or penis area where the tail ripper left off and continuing to about mid-chest level.

14. *First Codder* uses air knife to skin inside thigh of right hind leg, picking up at point where first legger left off [2].

15. *First Butter* uses air knife to separate hide from flesh around anus area.

16. *First Hock Vacuum* sticks enormous metal air vacuum over first hock (hind right leg, now clipped by first hock cutter) and holds it there for approximately seven seconds to clean area of fecal material and hair.

17. *First Hang Off* inserts metal hook attached to metal wheel pushed by "dogs" on separate overhead rail system into the hock hole created by the first hock cutter; guides/lifts the wheel onto the main rail track.

18. *Trimmer* uses handheld trimming knife to clear area between right hind leg and anus of fecal matter and hair; position exists primarily in winter months when there is more fecal matter on cattle.

19. *Unshackler/Low Raider* uses hands to release shackle as a machine called a low raider lowers left hind leg of cow. Cow now hangs from hook threaded through hole in leg created by the first hock cutter.

20. *Second Legger* uses hand knife to skin left hind leg, exactly as first legger has done for right hind leg [2].

21. *Second Hock Cutter* uses hock cutter to cut approximately six to eight inches of left hind leg off and then uses hand knife to poke hole between tendon and bone on left hind leg, exactly as first hock cutter has done for right hind leg.

22. *Second Codder* uses air knife to skin inside thigh of left hind leg, picking up where second legger left off [2].

23. *Second Butter* uses air knife to separate hide from flesh around anus area.

24. *Second Hock Vacuum* performs job identical to first

hock vacuum, except on left hind hock instead of right hind hock.

25. *Second Hang Off* takes hook attached to trolley and threads it through hole in left hind hock created by second hock cutter and then guides/pulls trolley wheel onto overhead rail. Cow is now suspended from two overhead trolleys, one hooked through each of its hind legs.

26. *Trimmer* trims area between anus and left hock for fecal matter and hair; position exists primarily in winter months when there is more fecal matter on cattle.

27. *Lower Belly Ripper* uses hand knife hook to open lower half of pattern in hide, picking up where upper belly ripper left off.

28. *Rim Over* uses air knife to skin front shoulder area of cattle, on both left and right sides.

29. *Right Flanker* uses air knife to skin right flank as far back as possible to prepare hide for side puller.

30. *Tail Bagger/Breed Stamper* places plastic bag over cow's tail, then fastens the bag on tail with rubber band to prevent feces from flying off tail onto rest of cow when the tail puller pulls hide off cow; stamps cattle according to breed—"A" for Angus, "H" for Hereford, and "C" for mixed Hereford and Angus—placing one mark on each side of rump.

31. *Bung Stuffer* wads up large sheet of thin paper and stuffs it deep into cow's anus to prevent feces from flying out of anus when the tail ripper pulls hide up off tail.

32. *Right Rumper* uses air knife to skin hide on right rump as far back as possible.

33. *Left Rumper* uses air knife to skin hide on left rump as far back as possible.

34. *Left Flanker* uses air knife to skin left flank as far back as possible, to prepare hide for side puller.

35. *Paper Liner* places thin piece of large rectangular paper on inside of each fold of flank hide to lessen likelihood of feces flying off outside and contaminating flesh during side puller operations.

36. *Ear/Nose Cutter* uses hand knife to slice off left ear and left nostril of cow. Stands on floor and wears full face shield to protect against copious amounts of blood.

37. *Ear/Nose/Horn Cutter* uses hand knife to slice off right ear and right nostril of cow. On any cows with horns, uses hydraulic shears and three and a half foot–long cleaver to sever horns. Stands on floor and wears full face shield to protect against copious amounts of blood.

38. *Bungee-Cord Attacher* uses hands to place bungee cord attached to metal clamps on hide, with paper between clamps. One end of bungee cord goes on trailing hide of one cow and is attached to leading hide of next cow to hold hide away from carcass to reduce amount of fecal material that is transferred to carcass by contact with hide.

39. *Third Hock Cutter* uses hydraulic shears to sever front left and right hoofs of cow, approximately five to eight inches up from bottom of each hoof. Severed hoofs are placed on conveyor that leads through hole in wall to foot cooking room.

40. *Side Puller Operator* operates large clamps attached to hydraulic arms by inserting flank hide into clamps, shutting clamps, then activating retraction on hydraulic arms to rip back hide and expose cow's front and belly [2].

40a. *Whizard Knife Belly Trimmer* uses circular Whizard knife to trim excess fat off belly area.

41. *Backer* uses air knives to cut pocket between hide and skin in middle back area of cattle while standing on moving conveyor. Backers must synchronize movements and work in tandem [3].

42. *Tail Puller Operator* uses hand to slide leading edge of back hide of each cow into long metal bar called banana bar. Once bar has slid all the way through the pocket created by the backer, the hydraulic-powered banana bar is activated and shoots upward, ripping hide off upper back area of cow, and ripping tail hide, including plastic bag on tail, off. At this point upper hide now hangs over lower half of cow, creating effect of white-fleshed upper body and lower body draped with hide.

43. *Bungee-Cord Remover* removes bungee cords and flank paper and places them on rack to be returned to bungee-cord attacher.

44. *Trimmer* uses hand knife to trim rump, anus, and tail area for fecal matter and hair; position exists primarily in winter when there is more fecal matter on cattle.

45. *Pizzle Remover* uses hand knife to sever penis or udder.

46. *Neck Opener* makes incision in hide at neck area, below incision made by first sticker.

47. *Down Puller Operator* uses hook to insert leading edge of draped hide into cylindrical metal spinner. Once inserted, worker uses levers to operate spinner, which rolls hide up as it spins, ripping hide off bottom half of carcass, including head. Once hide has been stripped completely from cow, spin is reversed and hide falls into collection chute. At this point, hide is completely removed from cow, and cow passes through wall which demarcates "clean" from "dirty."

48. *Dentition Worker* uses hook or fingers to open each cow's mouth and examine its teeth in order to determine whether it is over or under thirty months of age. If more than thirty months, cattle are at higher risk for BSE and dentition worker attaches red "30 month" tags to left side of cow's fore-

head and left shoulder area and plugs brain hole created by knocker with cork.

49. *Dentition Assistant* records number of each thirty-month cow on sheet of paper and attaches yellow tag to weasand from thirty-month cattle.

50. *Trimmer* uses hand knife to trim neck and back area of cattle for fecal material [2].

51. *Steam-Vacuum Worker* uses handheld metal vacuum nozzles to suction off any remaining fecal material or hair before cow enters prewash cabinet. Some workers vacuum inside first and second hock, others outside first and second hock, and others the back and gut areas [6].

52. *Fourth Hock Cutter* uses hock cutter to further trim two bottom hocks/shanks by about one or two inches to rid shanks of any contamination that might have occurred at down puller. Disposes of shank pieces in gray barrel.

53. *Ear Cutter* uses hand knife to cut ears completely off cow's head. Discards ears in gray barrel.

54. *Tail Cutter/Prewash Operator* uses hand knife to trim skinned tail. Also, turns off prewash cabinet when a thirty-month cow passes through. Prewash cabinet sprays liquid solution of lactic acid on cattle to reduce microbacterial contamination.

55. *Brisket Marker* uses hand knife to make incision in brisket, or front chest, area of cow in preparation for brisket saw operator.

56. *Brisket Saw Operator* uses electric saw to follow incision made by brisket marker and cut through brisket of cow, allowing access to gullet clearer.

57. *Head Dropper* uses hand knife to cut through neck, leaving head dangling by windpipe.

58. *Head Severer/Hanger* takes hook from head line, sinks it into base of skull, and severs windpipe so that head leaves body of cow and swings down onto head-line hook. Marks each head with tag corresponding to number of body it came from.

59. *Gullet Clearer* uses hand knife to reach into neck of cow and pull out gullet [2].

60. *Pregutter/Bladder Remover* uses hand knife to make incision in udder or penis area of cow, then reaches in with hand and knife and cuts out bladder. After puncturing bladder, discards it down narrow chute.

61. *Head Flusher* inserts double-headed nozzle into back of each head and sprays powerful stream of water through head to wash tongue and cheeks of any ingesta.

62. *Tongue Clipper and Washer* uses hydraulic shears to cut tongue bones, then uses hand knife to pull tongue out of head. Severs tongue and hangs it on hook next to head [2].

63. *Tongue Washer* sprays tongues with water to clean them.

64. *Pre-USDA Head Trimmer* uses hand knife to trim heads of any contamination before they reach USDA head-line inspectors.

65. *Post-USDA Head Trimmer* uses hand knife to trim head in accordance with USDA inspector directives.

66. *Gland Trimmer* removes heads from head-line hooks and uses hand knife to trim head glands before putting heads on chiseler table [3].

67. *Tongue Trimmer* removes tongues from head-line hooks and places them on trimming stand. Uses hand knife to trim bottom portion of tongue [2].

68. *Head Chiseler* uses pointed hydraulic metal chiseler operated by foot levers to scrape flesh off jaw bone, then uses

hydraulic metal bar, also activated by foot lever, to force jaw from skull. Sends both jaw and skull down small ramp to head table conveyor.

69. *Head, Cheek, and Lip Meat Trimmer* pulls jaws and skulls off head table conveyor and uses hand and/or Whizard knife to scrape head and cheek meat and to cut inner lips from skull and jaw. Uses hook or hand to toss meat into one of three chutes: head meat chute, cheek meat chute, or lip meat chute [5].

70. *Head, Cheek, and Lip Meat Boxer* operates trapdoors to head, cheek, and lip meat chutes in order to pack meat into boxes. Weighs boxes, using electronic scale, puts covers on boxes, and places boxes on offal conveyor.

71. *Bung Dropper* uses hand knife to cut around anus and cut large intestine away from anus. Pulls large intestine out of anus, caps it with plastic bag, then stuffs it back into anus to prevent feces from leaving large intestine when alimentary system is gutted and dropped onto viscera table [3].

72. *Weasand Rodder* uses hands to push a long rod into the windpipe of cow and pull out the weasand by separating it from windpipe. Clips end of weasand with serrated plastic clip.

73. *Tail Harvester* uses hand knife to sever tail and hang it on offal line.

74. *Gutter* steps onto moving viscera table and uses hand knife to slit cow open, connecting incisions made by pregutter and brisket saw operator. Reaches into incision with hand knife and severs heart, lungs, liver, pancreas, and gallbladder and drops these, along with stomach, small intestine, and large intestine, onto viscera table, where they are inspected by USDA inspectors. Widely considered one of the most physically challenging and skilled jobs on kill floor [5].

75. *Liver Harvester* uses hand knife to separate liver from gallbladder; cuts excess fat from liver and cuts gallbladder before dropping skin of gallbladder down gallbladder chute.

76. *Weasand Remover* uses hand knife and pulling motion to split weasand; severs weasand from windpipe and throws it back on viscera table.

77. *Offal Hanger* uses hands to hang hearts and weasand on offal line and livers on liver line; throws lungs, pancreases, windpipes, and any fetuses down pet-food chute.

78. *Paunch Separator* uses hand knife to separate paunch from intestines.

79. *Heart Trimmer* uses hand knife to open and trim hearts.

80. *Pet-Food Trimmer* uses hand knife to trim pet-food parts such as lungs, pancreas, and windpipe.

81. *Paunch Puller* uses handheld hooks to pull paunches and intestines off large viscera table onto small viscera table leading to gut room.

82. *Trimmer* uses hand knife to trim front shoulder area of cow for fecal material and hair.

83. *Split Saw Operator* stands on movable platform and holds large bandsaw. Uses foot to operate lever that mechanically raises and lowers platform while using bandsaw to split the cow in half through backbone, starting from anus down to neck [2].

84. *Spinal Cord Remover* uses vacuum and long tubular rod to remove spongy spinal cord from split cow [2].

85. *Kidney Dropper* uses hook to pull kidney down so that it hangs visibly in half-carcass.

86. *High Trimmer* uses hand knife and hook to trim hind hocks and rump area of cow prior to USDA trim-line inspection [3].

87. *Mid-Trimmer* uses hand knife and hook to trim middle area of cow prior to USDA trim-line inspection.

88. *Post-USDA Mid-Trimmer* uses hand knife and hook to correct any problems noted in middle area of the cow by USDA mid-level trim-rail inspector.

89. *Post-USDA High Trimmer* uses hand knife and hook to correct any problems noted in high area of cow by the USDA high-level trim-rail inspector.

90. *Kidney Remover* uses hand knife and hook to sever kidneys and throw them into gray barrel [2].

91. *Hot-Scale Operator* uses computer to weigh and print weight information on each cow. Uses pin to tag half-carcasses with slaughter number, lot number, sex, breed, and post-slaughter and evisceration weight.

92. *Outrail Trimmer* uses hand knife, hook, and electric well saw to trim cattle that failed USDA trim-rail inspection [2].

93. *Steam-Vacuum Worker* uses metal vacuum nozzle connected to suction tube to vacuum middle back area of half-carcass.

94. *Armpit Washer* uses high-pressure hose to clean armpit area of half-carcasses as they emerge from wash cabinet.

95. *Bone Crusher* uses hydraulically operated clamp machine to exert pressure on the ribs of half-carcasses, speeding cooling in the cooler.

96. *Internal Fat Cutter* uses hand knife to cut any remaining membranes or fat inside half-sides and score them with "X" [2].

97. *Thirty-Month Cattle Tagger/Recorder/Trimmer* places bright pink tags with number "30" on all thirty-month cattle to allow for easy identification by cooler workers. Records carcass number of each thirty-month cow to match against dentition assistant's records.

98. *Railer* works in cooler and uses hooks to organize incoming half-carcasses for overnight chilling [6].

99. *Liver Hanger* uses hands to take livers off liver line and hang on wheeled carts for chilling [2].

100. *Liver Packer* uses hands to take chilled livers off racks, wrap them inside plastic bags, pack them two to a box, and stack them on pallets to be shipped or stored [3].

101. *Tongue Washer and Packer* places tongues into circular metal washer and activates washer. After washing, wraps each tongue in shrink-wrap, boxes it, and places box on offal conveyor [2].

102. *Offal Packer* uses hands to take hearts, weasands, and tails off offal-line hooks. Tails are thrown into movable cart to be picked up by tail washer and packer. Weasands and hearts are thrown into separate boxes, placed on scales to be weighed, then boxed and put on offal conveyor belt.

103. *Tail Washer and Packer* pushes tail cart from offal packer to tail-washing machine. Places tails in tail washer then removes them, boxes them, and puts box on offal conveyor.

104. *Intestine and Paunch Separator* uses hand hook to push paunches down slide to paunch-opening room and intestines onto intestine conveyor. Spreads intestines so that they can be cleaned by water spray.

105. *Paunch Opener* uses hand knife to cut open stomachs and empty them of their partially digested contents. Pushes ingesta down chute, and hangs opened stomachs between two hooks on paunch line for transport to paunch-wash cabinet [4].

106. *Paunch Trimmer* uses hand knife to trim paunches after they have been washed in paunch-wash cabinet and puts them in circular vat washers to be washed again.

107. *Paunch Bandsaw Operator* uses bandsaw to cut open circular membranes about the size of a volleyball, then places

the membranes on conveyor leading to large metallic circular vat to be washed.

108. *Primary Intestine Washer* uses hands to remove large intestines from conveyor, thread them around coils located inside metal boxes, and activate coils, which release water to flush out large intestines. Once they are washed, removes intestines and places them in bin for secondary intestine washer [3].

109. *Secondary Intestine Washer* uses hands to pick up intestines from bin and thread them onto horizontal flusher. Activates flusher then removes intestines and puts them down slide to offal-packing room [2].

110. *Intestine Trimmer and Packer* trims intestines for any remaining fecal material and packs intestines in boxes. Places boxes on offal conveyor.

111. *Omasum and Tripe Washer and Refiner* uses hands to operate circular metal tripe and omasum washing vats [4].

112. *Omasum, Tripe, and Honeycomb Packer* uses hands and hand knife to pack boxes of washed and refined omasum, tripe, and honeycomb and place boxes on offal conveyor.

113. *Foot Cooker* uses hands to operate foot-cooking machines, which operate in spinning motion [2].

114. *Toenail Clipper* uses hands to stick each cooked foot into machine that operates by two serrated rollers. The rollers, activated by foot lever, spin together and break off "toenail," or tip of hoof, of each cow foot.

115. *Tendon Cutter* uses hands to push cow feet through band saw to cut off tendons on each foot. Tosses cut feet onto foot table.

116. *Foot Trimmer* uses hand knife to trim each foot of fat and discoloration [5].

117. *Foot Washer and Packer* sprays water on cow feet, packs six to a box, and places box on offal conveyor.

118. *Offal Packer* uses hands to pull boxes off conveyor, weigh them, print out sticker for each box, put sticker on box, activate machine that places plastic strap around the box, and put box back on conveyor. Opens boxes of hearts, head meat, and cheek meat, and puts one scoop of dry ice into each box [2].

119. *Supply-Room Staff* controls distribution of equipment to workers, including hairnets, gloves, safety gloves, aprons, knives, rubber bands, plastic bags, and hard hats.

120. *Kill Floor Sanitation Staff* cleans kill floor throughout day. Squeegees blood, dumps disposable meat and parts down discard chutes, takes pet-food parts to pet-food chute, and rinses down foam buildup [2].

121. *Nonproduction Sanitation and Laundry Staff* maintains bathroom and lunchroom areas and offices. Does laundry.

Appendix B: Cattle Body Parts and Their Uses

This appendix contains an outline of the uses of various cattle body parts other than meat. Referred to as offal on the kill floor and constituting up to 40 percent of a cow's weight, these cattle parts can generate anywhere from a tenth to a third of the income of an industrialized slaughterhouse. Some body parts—brains, kidneys, livers, pancreases, tails, thymuses, tongues, and udders—are sold directly as food. Others are further rendered and used in an astonishing array of fields and products, from medical research (fetal blood) to marshmallows (gelatin from bone) to cement additives (fat).[1]

Adrenal Glands: pharmaceutical uses, including cortisone, epinephrine, and norepinephrine

Arteries (carotid): human implantation as a femoropopliteal or iliofemoral substitute

Blood: sausage ingredient, sticking agent for insecticides, and blood meal for livestock and pet food

Bone: jellied products; refining sugars; soup; production

of gelatin for ice cream, mayonnaise, emulsion flavors, marshmallows, low-calorie sweeteners, wine, beer, and vinegar clarification, capsules and coatings for pharmaceutical pills, plasma expander for treatment of hemorrhages, trauma, and burns, photographic films and papers, bacteria culture media, smokeless gunpowder, foamer in fire extinguishers, and insecticide sprays; also used to produce phosphate fertilizer, bonemeal for livestock and pet food and glue for adhesive in plywood, furniture, veneer, paperboard, match heads, sandpaper, composition cork, mother-of-pearl, gummed tape, paper boxes, and bookbinding. Breastbones of young cattle used by plastic surgeons to replace facial bones.

Brain: cholesterol from brain used in vitamin D_3 synthesis, steroid pharmaceutical synthesis, and emulsifier in cosmetics

Cheek and Head Trimmings: sausage ingredient

Esophagus: sausage ingredient

Fat: oleomargarine, shortening, sweets, and chewing gum; tallow from beef fat used in industrial chemicals and synthetic oils, abrasives, shaving cream, asphalt tile, caulking compounds, cement additives, cleaners, cosmetics, deodorants, paints, polishes, perfumes, detergents, plastics, printing inks, synthetic rubber, and water-repellent compounds, lubricants, soap, candles, glycerin for medical uses and explosives (nitroglycerin), and meat meal for livestock and pet food

Feet: jelly

Fetal Blood and Serum: nutrient for tissue culture, research for vaccine production, cancer research, and virus propagation

Hair (tail, body, ear): paintbrushes, upholstery padding, felting, plaster retardant

Heart: processed lunchmeat, sausage ingredient

Hide: leather, rawhide, production of gelatin

Hide Trimmings: tankage, fertilizer, glue, inedible gelatin

Intestines: sausage casing

Liver: sausage ingredient

Oxtail: soup, stew

Skin trimmings: gelatin, jellied food

Skirt (wing of the diaphragm muscle): stew, sausage ingredient

Spleen: variety meat, pharmaceutical use to influence capillary permeability and treatment of blood and lymph diseases

Stomach (rumen, reticulum, abomasum): tripe, sausage ingredient

Weasand: sausage casing and ingredient

Notes

Chapter 1
Hidden in Plain Sight

1. Mark Kawar, "Freedom Is Fleeting for Cattle in Plant Escape: Cows Make a Stand," *Omaha World Herald,* August 5, 2004 (sunrise edition), 1–2. Without detectable irony, the story was placed immediately adjacent to an article about a recall of 497,000 pounds of beef contaminated with *E. coli.*

2. Slaughter statistics are compiled by the Food Safety Inspection Service and reported monthly by the National Agricultural Statistics Service of the United States Department of Agriculture (USDA). With the exception of declines during the Great Depression and the 1950s, these statistics tell the story of a steadily increasing consumption of meat in the United States throughout the twentieth century. In 1909 the per capita annual consumption of animal flesh was just over 120 pounds. In 2002 consumption per capita broke 200 pounds for the first time and has hovered near that since then. Translated into total number of pounds of dead animal consumed in the United States over the past century, the trend is staggering. In 1909 the approximately 90 million residents of the United States consumed 11.07 billion pounds of meat a year; in 2002 the population of meat eaters (about 93 percent) among the approximately 304 million U.S. residents consumed nearly 60 billion pounds of animal flesh.

These quantitative changes are reflected in the increasing concentration of the meat market. In 1970 the four biggest U.S. meatpackers controlled 21 percent of the beef market. Today only four corporations (Tyson Foods-IBP, ConAgra-Montfort, Cargill-Excel, and National Beef) control more than 80 percent of the U.S. market in beef, and the level of concentration is similar in other meat markets. Having acquired Iowa Beef Processors (IBP) in 2001, Tyson Foods is now the single largest processor and marketer of dead animals in the world, with control of 20 percent of the U.S. chicken market, 22 percent of the U.S. beef market, 20 percent of the U.S. pork market, and poultry operations on every continent in the world except Antarctica. Before its IBP acquisition, Tyson employed 120,000 people directly, not including an additional 7,038 independently contracted poultry growers raising chickens for Tyson's processing plants. In the fiscal year 2000 (again, before the company's acquisition of IBP) Tyson's total sales reached $23.8 billion. ConAgra's sales in 2001 netted $27.2 billion, $12.88 billion of which came directly from meat sales to the retail market. Today, a mere fourteen slaughterhouses account for 56 percent of all cattle killed; twelve slaughterhouses for 55 percent of all pigs killed; six slaughterhouses for 56 percent of all calves killed; and four slaughterhouses for 67 percent of all sheep and lambs killed.

This increase in economic concentration corresponds to enormous changes in the industry's structure and mode of operation. The "Big Five" conglomerate (Swift, Armour, Morris, Wilson, Cudahy) that had dominated the meat industry since the days of Upton Sinclair was replaced by a "new breed" of meatpackers relocated far away from urban areas in rural communities in Iowa, Colorado, Nebraska, Kansas, and throughout the South. The logic behind this relocation was twofold: first, to move closer to the supply of live animals; second, to take advantage of lower wages and weak unions to increase profits. The shift in geography also marked a significant shift in production techniques. The practice before the new breed of meatpackers was generally to ship carcasses whole to local butchers and grocery stores, where they would be further divided into edible portions. Today the practice of "boxed beef" has become the dominant industry model; instead of shipping whole carcasses, meatpackers do much of the cutting and portioning at processing plants, adding value to the meat as well as making it easier to transport.

The emphasis on speed in the new slaughterhouses—profits are measured by the quantity of processed meat per hour—translates into a rate of injury to workers that surpasses that of any other industry. Bureau of Labor Statistics data on job-related injuries and illnesses in the United States from 1976 to 2000 show that meatpacking has consistently topped the list as one of the most dangerous occupations in America. A person is more likely to become ill or injured working in a slaughterhouse than in shipbuilding and ship repairing, mining (although mining has more fatalities), construction work, working in an iron foundry, or any other private occupation typically thought of as dangerous.

3. For "a place that is no-place" see Noelie Vialles's brilliant account of French abattoirs in *Animal to Edible* (New York: Cambridge University Press, 1994), 15–32. For a history of the relocation of slaughterhouses from urban to rural areas, see Chris Philo, "Animals, Geography, and the City: Notes on Inclusions and Exclusions," in *Animal Geographies: Place, Politics, and Identity in the Nature-Culture Borderlands,* ed. Jennifer Wolch and Jody Emel (London: Verso, 1998). For a delightful history of the regulation of waste by the state that also touches on animal waste and animal slaughter, see Dominique Laporte, *A History of Shit,* trans. Radolphe el-Khoury (1978; Cambridge: MIT Press, 2000).

4. Zygmunt Bauman, *Modernity and the Holocaust* (Ithaca: Cornell University Press, 1989), 97; Pierre Bourdieu, *Language and Symbolic Power* (Cambridge: Polity, 1991), 207.

5. Mary Douglas, *Purity and Danger* (1966; New York: Routledge, 2002). The slaughterhouse has long been an archetypical metaphor in both popular and academic literature and media; the horror films alone that draw on slaughterhouse themes are too numerous to mention. For a recent academic discussion of the use of animal slaughter as an archetype of horror, see Talal Asad, *On Suicide Bombing* (New York: Columbia University Press, 2007).

6. HF 431, *The Iowa Legislature Bill Book,* 11–12, available at http://coolice.legis.state.ia.us/Cool-ICE/default.asp?Category=billinfo&Service=Billbook&menu=false&hbill=HF431 (accessed April 3, 2011).

Sections 9 and 10 of HF 589, the successor bill to HF 431, state:

"Sec. 9. *NEW SECTION.* 717A.2A Animal facility interference.

"1. A person is guilty of animal facility interference, if the person acts without the consent of the owner of an animal facility to willfully do any of the following:

"*a. (1)* Produce a record which reproduces an image or sound occurring at the animal facility as follows:

"(a) The record must be created by the person while at the animal facility.

"(b) The record must be a reproduction of a visual or audio experience occurring at the animal facility, including but not limited to a photographic or audio medium.

"*(2)* Possess or distribute a record which produces an image or sound occurring at the animal facility which was produced as provided in subparagraph (1).

"*(3)* Subparagraphs (1) and (2) do not apply to an animal shelter, a boarding kennel, a commercial kennel, a pet shop, or a pound, all as defined in section 162.2.

"*b.* Exercise control over the animal facility including an animal maintained at the animal facility or other property kept at the animal facility, with intent to deprive the animal facility of the animal or property.

"*c.* Enter onto the animal facility, or remain at the animal facility, if the person has notice that the facility is not open to the public. A person has notice that an animal facility is not open to the public if the person is provided notice before entering onto the facility, or the person refuses to immediately leave the facility after being informed to leave. The notice may be in the form of a written or verbal communication by the owner, a fence or other enclosure designed to exclude intruders or contain animals, or a sign posted which is reasonably likely to come to the attention of an intruder and which indicates that entry is forbidden.

"2. A person who commits the offense of animal facility interference is guilty of the following:

"*a.* For the first conviction, the person is guilty of an aggravated misdemeanor.

"*b.* For a second or subsequent conviction, the person is guilty of a class "D" felony.

"3. A person convicted of animal facility interference is subject to an order of restitution as provided in chapter 910.

"Sec. 10. *NEW SECTION.* 717A.2B Animal facility fraud.

"1. A person is guilty of animal facility fraud, if the person willfully does any of the following:

"*a.* Obtains access to an animal facility by false pretenses for the purpose of committing an act not authorized by the owner of the animal facility.

"*b.* Makes a false statement or representation as part of an application to be employed at the animal facility, if the person knows the statement to be false, and makes the statement with an intent to commit an act not authorized by the owner of the animal facility.

"2. A person who commits the offense of animal facility fraud is guilty of the following:

"*a.* For the first conviction, the person is guilty of an aggravated misdemeanor.

"*b.* For a second or subsequent conviction, the person is guilty of a class "D" felony.

"3. A person convicted of animal facility fraud is subject to an order of restitution as provided in chapter 910.

"4. This section does not apply to an animal shelter, a boarding kennel, a commercial kennel, a pet shop, or a pound, all as defined in section 162.2" (HF 589, *The Iowa Legislature Bill Book,* 4–5, available online at http://

coolice.legis.state.ia.us/Cool-ICE/default.asp?Category=billinfo&Service=Billbook&menu=false&ga=84&hbill=HF589 (accessed April 3, 2011).

At the time this book went to press, HF 589, already passed by the Iowa House of Representatives, had also passed the Iowa Senate Committee on Agriculture with a recommendation for adoption and was awaiting debate on the senate floor. A number of intriguing amendments, including an amendment to extend the "Interference" and "Fraud" language in the bill to pregnancy termination clinics and medical facilities, had also been proposed by Iowa senator Matt McCoy.

7. HF 431, 12.

8. HF 589, 5.

9. HF 589 expansively defines "record" as "any printed, inscribed, visual, or audio information that is placed or stored on a tangible medium, and that may be accessed in a perceivable form, including but not limited to any paper or electronic format" (2).

10. See George Orwell's classic essay "Politics and the English Language" in *Shooting an Elephant and Other Essays* (New York: Harcourt, 1946).

11. Zygmunt Bauman, "The Phenomenon of Norbert Elias," *Sociology* 13, no. 1 (1979): 122; Norbert Elias, *The Civilizing Process*, trans. Edmund Jephcott (Malden, Mass.: Blackwell, 2000), 103.

12. Elias, *Civilizing Process*, 102. The frontiers of repugnance—that is, the reach of a map of phenomena that provoke both moral and physical disgust—expand with the civilizing process, a term Elias uses broadly to refer to "the process of advancing integration, increased differentiation of social functions and interdependence, and the formation of ever-larger units of integration" (254). Writing of the movement from small and middle-sized lordships to larger feudal lordships to kingdoms and then to the modern state—that is, of the increasing monopolization and concentration of violence in a centralized authority and its increasingly systematic cleansing from the realms of everyday life—Elias notes: "With each shift, the network of dependencies intersecting the individual has grown larger and changed in structure; and with each shift, the molding of behavior and of the whole emotional life, the personality structure, is also changed" (254). What is remarkable about Elias's account is not his history of increasing economic interdependencies and the monopolization of violence by the territorial political state—this is heavily traversed terrain—but rather his singular connection of this macro-institutional account to an argument about changes in the psychosocial and emotional makeup of individual humans.

In his own account of the olfactory ordering of the social order, Alain Corbin reinforces Elias's history of civilization as a process of quarantine and concealment, showing how a concern for classifying, confining, and segregating by smell in late-eighteenth-century Paris provided the impetus for what later came to be interpreted primarily as a reordering of the visual economy. Concerned with "the odors of the sick town," these pre-Pasteurian social reformers brought strategies of segregation and quarantine to bear on a host of miasmal dangers, including dungeons, prisons, tombs, and slaughterhouses. Of slaughterhouses in particular, Corbin writes, "The presence of slaughterhouses within towns promoted indignation and intensified vigilance concerning decaying carcasses. The urban slaughterhouse was an amalgam of stenches. In butchers' narrow courtyards odors of dung, fresh refuse, and organic remains combined with foul-smelling gases escaping from intestines. Blood trickled out into the open air, ran down the streets, coating the paving stones with brownish glazes, and decomposed in the gaps. Because blood transmitted 'fixed air,' it was the most eminently putrescent of animal remains. The malodorous vapors that impregnated roadways and traders' stalls were some of the deadliest and the most revolting; they 'make the whole body susceptible to putridity.' Often the stifling odors of melting tallow added to this foul-smelling potpourri" (Corbin, *The Foul and the Fragrant: Odor and the French Social Imagination* [Cambridge: Harvard University Press, 1986], 31).

13. Michel Foucault, "Two Lectures," in *Power/Knowledge: Selected Interviews and Other Writings*, ed. Colin Gordon (New York: Pantheon), 105.

14. Michel Foucault, "The Eye of Power," in *Power/Knowledge*, 147, 158. See also Michel Foucault, *Discipline and Punish: The Birth of the Prison*, trans. Alan Sheridan (New York: Vintage, 1977). Note that the ideal of "total and circulating mistrust" becomes problematic in the actual operation of the Panopticon, opening possibilities for breakdown and resistance. Key among these is the problem of the Original Overseer, the eye of God. As Foucault's interlocutor Michelle Perrot notes, "The working of the Panopticon is somewhat contradictory from this point of view. There is the chief inspector who watches over the prisoners from the central tower; but he watches his subordinates as well, the personnel in the hierarchy . . . an unbroken succession of observations recalling the motto: each comrade becomes an overseer. Yet it's Bentham who begins by relying on a single power, that of the central tower. As one reads him one wonders who he is putting in the tower. Is it the eye of God?" (In Foucault, "Eye of Power," 156–157).

15. James C. Scott, *Seeing Like a State: How Certain Schemes to Improve the Modern Condition Have Failed* (New Haven: Yale University Press, 1998). For Scott, as for Foucault, the pursuit of visibility is always just that: a pursuit. Never perfectly realized, and far from totalizing, this pursuit always

offers points of resistance, sometimes generates self-defeating pathologies of its own making, and must continually reinvent itself with new strategies and tactics.

16. James C. Scott, *The Art of Not Being Governed: An Anarchist History of Highland Southeast Asia* (New Haven: Yale University Press, 2009), xii.

17. Once I had decided on industrialized slaughterhouses as a research site, I began by ruling out poultry slaughterhouses because the physical uniformity of chickens allows for an automated killing process in which humans are largely absent. This left either hog or cattle slaughterhouses as possible research sites. My next step was to compile data from the Food Safety Inspection Service of the USDA, which publishes a directory of all federally inspected slaughterhouses in the United States. Since the focus of my study was on large-scale industrialized slaughterhouses, I limited my search to those employing five hundred workers or more.

Using these data, I mapped the geographic location of each of these slaughterhouses and isolated specific areas where they are clustered. Because no guarantee existed that I could gain access to a particular slaughterhouse (or to any slaughterhouse at all, for that matter), I sought a location with a number of slaughterhouses within a small geographical area. Additionally, given my interest in the production of social invisibility—in how the slaughterhouse is constructed as a place that is "no-place"—I wanted to find a slaughterhouse in an urban location. Choosing a remote rural location might offer an easy geographical answer to the question that would not necessarily be improved or amended by additional participant-observation work. Slaughterhouses are socially invisible, it could be argued, because they have for the most part moved to rural locations out of the sight, sound, and smell of the vast majority of meat consumers. By choosing an urban or semi-urban location, I was able to examine some of the tensions and paradoxes that arise when the signs of an activity that is meant to be invisible are exhibited in plain sight.

Filtering the USDA data on industrialized meat production plants employing five hundred or more persons through these criteria, I identified Omaha as a potentially fertile location for my field research. With a population of approximately 390,000, Omaha is one of the fifty largest urban areas in the United States. Within the city limits of Omaha and Council Bluffs, Iowa (which lies just across the Missouri River), there are several industrialized slaughter and fabrication operations, employing more than six thousand workers. Additionally, a number of industrialized slaughterhouses are located in several smaller towns in the Omaha area. In terms of density of slaughterhouses within and around a major urban area, Omaha appeared to be a location with exceptional promise. In subsequent chapters I outline in greater detail the story of my access to the particular slaughterhouse I worked in.

18. For networks of power, see Michael Burawoy, "The Extended Case Method," *Sociological Theory* 16, no. 1 (1998): 4–33. For "webs of local association" see John Van Maanen, "Playing Back the Tape," in *Experiencing Fieldwork,* ed. William B. Shaffir and Robert A. Stebbins (London: SAGE, 1991), 40. For the position of the researcher and its relation to knowledge production, see Samer Shehata, "Ethnography, Identity, and the Production of Knowledge," in *Interpretation and Method: Empirical Research Methods and the Interpretive Turn,* ed. Dvora Yanow and Peregrine Schwartz-Shea (Armonk, N.Y.: Sharpe, 2006), 244–263. For a methodological reflection on political ethnography that draws on my research, see Timothy Pachirat, "The Political in Political Ethnography: Dispatches from the Kill Floor," in *Political Ethnography; What Immersion Contributes to the Study of Power,* ed. Edward Schatz (Chicago: University of Chicago Press, 2009), 143–162. On the importance of "place-ness" to the trustworthiness of interpretive political ethnography, see Dvora Yanow, "Dear Author/Dear Reviewer," in *Political Ethnography,* 275–302. For information-based case-selection strategy, see Bent Flyvbjerg, "The Power of Example," in his *Making Social Science Matter* (Cambridge: Cambridge University Press, 2001), 66–87.

19. Decisions about how to access the research site can have major implications for the perspectives and information available to the researcher during the course of the research. Even after defining my field of research and narrowing it down to a potential geographic location, I still faced a series of important decisions about access. Should I attempt access as a researcher, writing a letter of inquiry to the relevant officials at various slaughterhouses requesting to be allowed into their plants? Should I attempt a kind of proxy access through interviews or surveys of slaughterhouse workers, relying on selectively highlighted information from their accounts to reconstruct a world I would not enter directly myself? Or should I attempt access as an entry-level worker, showing up in person to apply at a plant without notifying anyone of my interests as a researcher? How I negotiated these choices would profoundly affect my line of sight within the slaughterhouse. Would it be a vision from above, from below, or from the outside looking in? And, most important, what mode of seeing was best suited to addressing the concerns animating my research?

Having judged that a strategy of formal access was likely to fail and prove inadequate even if successful, and decided against relying exclusively on proxy access through interviews with workers, I decided to attempt to gain access as an entry-level employee.

In considering this option, I had a large sociological literature of the workplace to draw on, including Donald Roy, "Restriction of Output in a Piecework Machine Shop" (Ph.D. diss., University of Chicago, 1952); Michael Burawoy, *The Colour of Class: From African Advancement to Zambianization* (Manchester: Manchester University Press for Institute for Social Research, 1972); Richard Pfeffer, *Working for Capitalism* (New York: Columbia University Press, 1979); Robert Linhart, *The Assembly Line*, trans. Margaret Crosland (Amherst: University of Massachusetts Press, 1981); Ruth Cavendish, *Women on the Line* (Boston: Routledge and Kegan Paul, 1982); Maria Fernandez-Kelly, *For We Are Sold, I and My People: Women and Industry in Mexico's Frontier* (Albany: State University of New York Press, 1983); Tom Juravich, *Chaos on the Shop Floor: A Worker's View of Quality, Productivity, and Management* (Philadelphia: Temple University Press, 1985); Louise Lamphere, "Bringing the Family to Work: Women's Culture on the Shop Floor," *Feminist Studies* 11 (1985): 519–540; Laurie Graham, *On the Line at Subaru-Isuzu: The Japanese Model and the American Worker* (Ithaca: ILR Press, 1994); and Barbara Ehrenreich, *Nickel and Dimed: On (Not) Getting By in America* (New York: Holt, 2001). These studies share a strategy of gaining access to a workplace without notifying employers and fellow employees (at least initially) of the research intention motivating the employment. In many cases, the researchers understand covert employment to be the only means of entry to the workplace; disclosure of the research intention would have barred the researcher from access. As the sociologist Michael Burawoy writes, "To penetrate the shields of the powerful the social scientist has to be lucky and/or devious" ("Extended Case Method," 22).

In addition to these studies, there have been three ethnographic studies of slaughterhouses that have relied on covert access: William E. Thompson, "Hanging Tongues: A Sociological Encounter with the Assembly Line," *Qualitative Sociology* 6, no. 3 (1983): 215–237; Deborah Fink, *Cutting into the Meatpacking Line* (Chapel Hill: University of North Carolina Press, 1998); and Steve Striffler, "Inside a Poultry Plant," in *Chicken: The Dangerous Transformation of America's Favorite Food* (New Haven: Yale University Press, 2007). Contrasting these studies with accounts of meatpacking work in which the researcher formally requested, but was denied, access to the kill floor by management—see, for example, Donald Stull and Michael Broadway, eds., *Slaughterhouse Blues: The Meat and Poultry Industry in North America* (Belmont, Calif.: Thomson/Wadsworth, 2004), and Donald Stull,

"Knock 'em Dead: Work on the Killfloor of a Modern Beefpacking Plant," in *Newcomers in the Workplace: Immigrants and the Restructuring of the U.S. Economy*, ed. Louise Lamphere, Alex Stepick, and Guillermo Grenier (Philadelphia: Temple University Press, 1994)—or where the researcher worked on the kill floor with management approval and submitted his findings to management (Ken C. Erickson, "Guys in White Hats: Short-Term Participant Observation Among Beef-Processing Workers and Managers," in *Newcomers in the Workplace*, 96) persuaded me that requesting the formal permission of management would severely limit my access in the plant if it was granted at all.

This literature convinced me that gaining covert access to the kill floor as an entry-level employee was both possible as a practical matter and promising as a means of achieving the kind of ethnographic, empathetic understanding that drove my research interests. Still, I understood clearly that the use of deception to gain access represented a potential cost as well as an opportunity. Rather than deductively deciding ex ante whether the use of deception in research is always or never justified, a better course of action is a context-specific examination of the purposes for using deception in each particular case. The ethics of concealment are not the only ethics to be considered in participant-observation work; the ethics of power are also relevant. In the context of my study, I knew beforehand that I would be entering a research terrain with an extreme imbalance of power. There would be no neutral mode of access. To enter with the full permission of slaughterhouse management would be to enter on the side of management. Although this mode of access would allow me to be "ethical" in the sense of minimizing the use of deception over the fact that I was conducting research, it could nonetheless potentially implicate me ethically in terms of how I positioned myself in the power hierarchy within the plant. On the other hand, to enter as an entry-level worker without management's knowledge of my research agenda would allow me to approximate more closely the experiences of the slaughterhouse from below despite the fact that it would require me to deceive management about my purpose for being there. I judged the latter tradeoff to be the preferable ethical decision in a situation bereft of a single ethically pure approach.

The second response to the concerns raised by covert research is more practical in nature. In conjunction with an oversight committee, I developed a set of guidelines to address concerns about the protection of the people I would meet in the course of my research. Together we decided that all information identifying specific companies and workers at any level of the company would be altered in order to protect anonymity. The goal of the research was never to expose any one company or set of persons but rather to create a portrait of slaughterhouse work as it might take place in any number of industrialized slaughterhouses. Further, fieldnotes which might compromise myself or co-workers would be stored at a secure location away from my residence in the event that my identity was compromised and my residence searched. Finally, after I left the plant, any further contact with former co-workers or other slaughterhouse workers in the form of conversations and interviews would be predicated on a full disclosure of my status as a researcher.

20. Anna Tsing, *In the Realm of the Diamond Queen* (Princeton: Princeton University Press, 1993), 33. These are all components of what Bent Flyvbjerg describes as "the irreducible quality of good case narrative" in *Making Social Science Matter* (Cambridge: Cambridge University Press, 2001), 84. For two radically contrasting exemplars of this kind of close, studied approach, see Harold C. Conklin, *Hanunóo Agriculture: A Report on an Integral System of Shifting Cultivation in the Philippines* (Rome: Food and Agriculture Organization of the United States, 1957), and James Agee and Walker Evans, *Let Us Now Praise Famous Men* (1941; Boston: Houghton Mifflin, 2001). Of course, the "thick narrative" accounts of killing work I offer in this study also have been pieced together and selectively edited from thousands of pages of fieldnotes. Only a fraction of the conversations, incidents, descriptions, and interactions that I recorded in the course of my fieldwork make an appearance here. To suggest or imply otherwise would be to fall into the fallacy of implying that an authorless, unmediated transference of the fieldwork experience to the reader is taking place, a version of Donna Haraway's "god trick" in which social scientists contemplate with fascination their own absence. Fieldnotes are themselves highly stylized, selective accounts written from particular, embodied vantage points that are characterized by partial rather than unrestricted fields of vision. For the "god trick," see Donna Jeanne Haraway, *Simians, Cyborgs, and Women: The Reinvention of Nature* (London: Free Association Books, 1991), 191. On fieldnotes, see, for example, Robert M. Emerson, Rachel Fretz, and Linda Shaw, *Writing Ethnographic Fieldnotes* (Chicago: University of Chicago Press, 1995). For a nuanced historical and theoretical account of the construction of ethnographic authority see James Clifford, "On Ethnographic Authority," *Representations* 2 (Spring, 1983): 118–146.

21. Henry Miller, "Reflections on Writing," in *Wisdom of the Heart* (New York: New Directions, 1941), 27, also quoted in Flyvbjerg, *Making Social Science Matter,* 133, and C. Roland Christensen with Abby J. Hansen, eds., *Teaching and the Case Method* (Boston: Harvard Business School Press, 1987), 18. Ian Miller, *The Anatomy of Disgust* (Cambridge: Harvard Univer-

sity Press, 1997), 22. Miller notes of the relation between disgust and sensory detail: "Disgust differs from other emotions by having a unique aversive style. The idiom of disgust consistently invokes the sensory experience of what it feels like to be put in danger by the disgusting, of what it feels like to be close to it, to have to smell it, see it, or touch it. Disgust uses images of sensation or suggests the sensory merely by describing the disgusting thing so as to capture what makes it disgusting. Images of sense are indispensable to the task. We thus talk of how our senses are offended, of stenches that make us retch, of tactile sensations of slime, ooze, and wriggly, slithering, creepy things that make us cringe and recoil. No other emotion, not even hatred, paints its object so unflatteringly, because no other emotion forces such concrete sensual descriptions of its object" (9).

Chapter 2
The Place Where Blood Flows

1. Interview with long-time South Omaha resident, March 9, 2004.

2. For a history of slaughterhouses in Omaha specifically, see Gail DiDonato, "Building the Meatpacking Industry in South Omaha, 1883–1898" (master's thesis, University of Nebraska, Omaha, 1989). Anna Williams offers a fascinating history of slaughterhouses in the United States from the perspective of the urban visual economy in "Nothing but Bodies: Nineteenth Century Representations of Animals in Georges Cuvier's Natural System and U.S. Industrial Meat Production" (Ph.D. diss., University of Rochester, 2000). For general histories of the meatpacking industry in the United States, see Rudolf Clemen, *The American Livestock and Meat Industry* (New York: Ronald Press, 1923); James Barrett, "Work and Community in *The Jungle:* Chicago's Packinghouse Workers, 1894–1922" (Ph.D. diss., University of Pittsburgh, 1981); Rick Halpern, *Down on the Killing Floor: Black and White Workers in Chicago's Packinghouses, 1904–1954* (Chicago: University of Illinois Press, 1997); Rick Halpern and Roger Horowitz, *Meatpackers: An Oral History of Black Packinghouse Workers and Their Struggle for Racial and Economic Equality* (New York: Twayne, 1996); Jimmy M. Skaggs, *Prime Cut: Livestock Raising and Meatpacking in the United States, 1607–1983* (College Station: Texas A&M University Press, 1986); William Cronin, "Annihilating Space: Meat," in his *Nature's Metropolis: Chicago and the Great West* (New York: Norton, 1991), 207–247; Margaret Walsh, *The Rise of the Midwestern Meatpacking Industry* (Lexington: University of Kentucky Press, 1982); and Wilson J. Warren, *Tied to the Great Packing Machine: The Midwest and Meatpacking* (Iowa City: University of Iowa Press, 2007).

3. "The Big Man," in *Flogger Songs,* ed. Lowell Otte (Cedar Rapids: Torch, 1926). Collected from the workers in South Omaha's stockyards, these songs, writes Otte, "have been scribbled where and when you will: in the chute-house, at eleven-spot scale house, on fences, and on stone slabs. They have been composed while driving cattle, while yarding hogs, on the 'bum,' and during 'bull-sessions' with fellow floggers. They have their own vocabulary; their lapses from English would make the author's old rhetoric instructor turn gray. But they are sincere, so their author places them in your hands with only that sincerity to commend them."

4. For a brilliant history of aerial bombardment that gives special attention to the idea of distance, see Sven Lindqvist, *A History of Bombing,* trans. Linda Haverty Rugg (New York: New Press, 2001).

5. See George Orwell, "Politics and the English Language," in *Shooting an Elephant and Other Essays* (New York: Harcourt, 1946); Murray Edelman, "Political Language and Political Reality," *PS: Political Science and Politics* 18, no. 1 (1985): 10–19; and Keith Allan and Kate Burridge, *Euphemism and Dysphemism: Language Used as Shield and Weapon* (New York: Oxford University Press, 1991).

Chapter 3
Kill Floor

1. For an enlightening exploration of euphemism and dysphemism, see Keith Allan and Kate Burridge, *Euphemism and Dysphemism: Language Used as Shield and Weapon* (New York: Oxford University Press, 1991). Their account emphasizes the face-saving aspects of language, however, and touches only briefly on the deceptive uses of euphemism.

2. In addition to my own observations, these instances are documented in publicly accessible animal-handling Noncompliance Reports filed by USDA inspectors who observed sentient cattle being cut into after passing through the electrical stimulation and bleed pit area. Noncompliance Reports can be accessed through the Food Safety and Inspection Service division of the United States Department of Agriculture.

Chapter 4
"Es todo por hoy"

1. The role of family in fieldwork, particularly in the kind of fieldwork that I conducted, is one that I regretfully leave aside in this book.

2. My fear is similar to that expressed by Maria Patricia Fernandez-Kelly when she covertly applied to work in several maquiladoras (border factories) in Ciudad Juarez: "At the same time a doubt struck me. Would several years of penmanship practice at a private school in Mexico City and flawless spelling give motive for suspicion? When trying to modify a writing style and idiosyncrasies one comes face to face with the meaning of participant observation, its pitfalls and limitations." Maria Patricia Fernandez-Kelly, *For We Are Sold, I and My People* (Albany: State University of New York Press, 1983), 113.

3. Italo Calvino, *Invisible Cities*, trans. William Weaver (New York: Harcourt, 1979), 89.

Chapter 5
One Hundred Thousand Livers

1. Claude Pavaux, *Color Atlas of Bovine Visceral Anatomy*, scientific adaptation by G. C. Skerritt, English translation by M. N. Samhoun (London: Wolfe Medical Publications, 1983).

2. Wright Morris, *The Home Place* (1948; Lincoln: University of Nebraska Press, 1999), 76.

Chapter 6
Killing at Close Range

1. John Lachs, *Intermediate Man* (Indianapolis: Hackett, 1985), 13.

Chapter 7
Control of Quality

1. These tasks consist of 1) placing gray plastic waste barrels mounted on wheels at different locations on the kill floor for condemned meat, which will later be sent to rendering; 2) setting up the plastic bags and rubber bands at the bung dropper stations (circle 71 in figure 2); 3) pouring indelible ink into the inkwells at USDA line inspector stations; 4) filling a bin near the weasand-rodding station with special serrated plastic clips (circle 72); 5) bringing two thermometers from the kill floor office to a maintenance worker, who attaches them to the 185 cabinet; and 6) placing clipboards with

paperwork for documentation at four locations on the kill floor. Failure to perform any one of these tasks satisfactorily can lead to an NR.

2. According to USDA regulations, this paperwork must be kept on file by the plant for a year and can be requested by a USDA inspector at any time.

3. A key reason why the kill floor must have at least two QCs is the random timing of the hourly tests at the CCP-1 stand and the CCP-2 table. If the randomly generated numbers call for simultaneous or closely spaced CCP-1 and CCP-2 tests, there must be more than one QC available to perform them.

4. The probability of receiving a second NR can be lowered if the QC inspects each and every piece of meat and weasand in the boxes randomly chosen for the CCP-2 test. It takes between fifteen and twenty minutes (each hour) to inspect a sample of meat from each box, however, and if the entire contents of each box were to be inspected, the time spent on CCP-2 tests would increase to between thirty and forty minutes an hour, leaving the QC little time for the other quality-control tasks.

5. Jill's rationalization about not being the last point in the inspection line clearly helps ease her conscience about being responsible for making someone ill because of contamination on carcasses that she has failed to flag. Part of the problem with this rationalization, however, is that the fabrication department has its own written HACCP plan, which operates on the assumption that the raw material entering the department is free of contamination. Although it is true that one fabrication worker is assigned to trim the carcasses as they enter the floor, this trimmer looks only at the upper side of the carcasses for rail dust and grease that might have fallen onto the carcasses as they hung in the cooler; there is no one looking for fecal contamination and no worker examining the lower half of the carcass. Here, again, because of the linear process of production in the slaughterhouse, the failure to document contamination during any CCP test threatens the food supply at all subsequent points in the production process not only for the specific contaminated carcass but for the entire population of carcasses for which it was a sample.

Chapter 8
Quality of Control

1. For a discussion of hidden transcripts and backstage spaces see James C. Scott, *Domination and the Arts of Resistance* (New Haven: Yale Uni-

versity Press, 1992), and Erving Goffman, *The Presentation of Self in Everyday Life* (New York: Anchor, 1959).

2. Michel Foucault, "The Eye of Power," in *Power/Knowledge: Selected Interviews and Other Writings*, ed. Colin Gordon (New York: Pantheon), 158. I discuss Foucault's theory in Chapter 1.

3. For a unique treatment of the animal voice in the industrialized slaughterhouse, see Mick Smith, "The 'Ethical' Space of the Abattoir: On the (In)human(e) Slaughter of Other Animals," *Human Ecology Review* 9, no. 2 (2002): 49–58.

Chapter 9
A Politics of Sight

1. After leaving the slaughterhouse, I stayed in Omaha for another eighteen months, conducting, on a much less grueling schedule, participant-observation research and interviews with community and union organizers, slaughterhouse workers, USDA inspectors, cattle ranchers, and small-slaughterhouse operators.

2. Georges Bataille, "Slaughterhouse," trans. Paul Hegarty, in *Rethinking Architecture: A Reader in Cultural Theory*, ed. Neil Leach (New York: Routledge, 1997), 22. In the same place, Bataille also writes, "In fact, the victims of this curse are not butchers or animals, but the good people themselves who, through this, are only able to bear their own ugliness, an ugliness that is effectively an answer to an unhealthy need for cleanliness, for a bilious small-mindedness, and for boredom. The curse (which terrifies only those who utter it) leads them to vegetate as far as possible from the slaughterhouses. They exile themselves, by way of antidote, in an amorphous world, where there is no longer anything terrible, and where, enduring the ineradicable obsession with ignominy, they are reduced to eating cheese. The slaughterhouse emerges from religion insofar as the temples of times past (not to mention the Hindu temples of today) had a dual purpose, being used for both supplication and slaughter. From this, without doubt (and this much can be adjudged from the chaotic appearances of the abattoirs of today), comes the startling coincidence of the mythological mysteries with the lugubrious grandeur that characterizes the places where blood flows. It is curious to see an aching regret being expressed in America: W.B. Seabrook finds that current customs are insipid, remarking that the blood of sacrifice is not mixed in with cocktails."

3. Michel Foucault, "The Eye of Power," in *Power/Knowledge: Selected Interviews and Other Writings*, ed. Colin Gordon (New York: Pantheon), 158.

4. Ibid., 152–153, 152.

5. Ursula Le Guin, *The Dispossessed: An Ambiguous Utopia* (New York: Harper and Row, 1974), 132.

6. Ibid., 98.

7. Michael Pollan, "An Animal's Place," *New York Times Sunday Magazine,* November 10, 2002. The open-air chicken abattoir is also discussed in Pollan's *The Omnivore's Dilemma: A Natural History of Four Meals* (London: Penguin, 2006), 226–238.

8. HF 589, *The Iowa Legislature Bill Book,* 5, available online at http://coolice.legis.state.ia.us/Cool-ICE/default.asp?Category=billinfo&Service=Billbook&menu=false&ga=84&hbill=HF589 (accessed April 3, 2011).

9. Jean-Jacques Rousseau, *The First and Second Discourses,* trans and ed. Victor Gourevitch (New York: Harper and Row, 1990), 160–163. The passage immediately preceding this is equally evocative in its description of pity in animals: "I speak of Pity, a disposition suited to beings as weak and as subject to so many ills as we are; a virtue all the more universal and useful to man as it precedes the exercise of all reflection in him, and so Natural that the Beasts themselves sometimes show evident signs of it. To say nothing of the tenderness Mothers feel for their young and of the dangers they brave to protect them, one daily sees in the repugnance of Horses to trample a living Body underfoot; an animal never goes past a dead animal of its own Species without some restlessness: Some even give them a kind of burial; and the mournful lowing of Cattle entering a Slaughter-House conveys their impression of the horrible sight that strikes them."

10. Anthony Giddens, *Modernity and Self-Identity* (Cambridge: Polity, 1991), esp. chaps. 2 and 5. Along with death, Giddens includes illness, mental disorders, and sexuality as experiences that have the potential to raise existential anxieties that have been systematically sequestered from modern life. Hannah Arendt, *Eichmann in Jerusalem* (New York: Viking, 1963), 106. Max Horkheimer, "Materialismus und Moral" in *Zeitschrift für Sozialforschung* 2, no. 2 (1933): 184. The context of Horkheimer's phrase is discussed by Seyla Benhabib in *Critique, Norm, and Utopia* (New York: Columbia University Press, 1986), 200: "Morality belongs to a certain form which human relations have assumed on the basis of the economic organization of the bourgeois order. With the transformation of these relations through their rational regulation, morality steps into the background. Humans can then fight together against their own suffering and sickness . . . but in nature, however, misery and death reign. Human solidarity is nonetheless an aspect of the solidarity of the living in general. Progress in the realization of the first will strengthen our sense for the latter. Animals need humans." Horkheimer's signaling of the dependence of progress among the solidarity

of the living on progress in human solidarity is germane for the double-edged nature of slaughterhouse violence against humans and cattle. Lev Tolstoy, *War and Peace,* trans. Constance Garnett (New York: Modern Library, 1994), 1224.

11. Norbert Elias, *The Civilizing Process,* trans. Edmund Jephcott (Malden, Mass.: Blackwell, 2000), 171; Yi-Fu Tuan, *Dominance and Affection: The Making of Pets* (New Haven: Yale University Press, 1984), 89–90.

12. Susan Sontag, *Regarding the Pain of Others* (New York: Farrar, Straus, and Giroux, 2003), 81. For excellent work in this direction that draws on a careful reading of Adam Smith's *The Theory of Moral Sentiments,* see Luc Bultanski, *Distant Suffering: Morality, Media, and Politics,* trans. Graham Burcell (New York: Cambridge University Press, 1999).

13. Sontag, *Regarding the Pain of Others,* 81.

Appendix B
Cattle Body Parts and Their Uses

1. The information in this appendix was synthesized from the following sources: H. W. Ockerman and C. L. Hansen, *Animal By-Product Processing and Utilization* (Lancaster, U.K.: Technomic, 2000); A. M. Pearson and T. R. Dutson, eds., *Inedible Meat By-Products: Advances in Meat Research,* vol. 8 (London: Elsevier, 1992); Don Franco and Winfield Swanson, eds., *The Original Recyclers* (Alexandria, Va.: National Renderers Association, 1996); and David L. Meeker, ed., *Essential Rendering* (Alexandria, Va.: National Renderers Association, 2006).

Index